Springer Series in Statistics

Springer Series in Statistics

(continued after index)

Frédéric Ferraty
Philippe Vieu

Nonparametric Functional Data Analysis

Theory and Practice

With 29 Illustrations

 Springer

Frédéric Ferraty
Laboratoire de Statistique et
    Probabilités
Université Paul Sabatier
Toulouse CX31062
France
ferraty@cict.fr

Philippe Vieu
Laboratoire de Statistique et
    Probabilités
Université Paul Sabatier
Toulouse CX31062
France
vieu@cict.fr.

Library of Congress Control Number: 2005936519

ISBN-10: 0-387-30369-3   Printed on acid-free paper.
ISBN-13: 978-0387-30369-7

Printed in the United States of America.   (MVY)

9 8 7 6 5 4 3 2 1

springer.com

To the working group STAPH

Preface

This work is the fruit of recent advances concerning both nonparametric statistical modelling and functional variables and is based on various publications in international statistical reviews, several post-graduate courses and international conferences, which are the result of several years of research. In addition, all these developments have their roots in the recent infatuation for functional statistics. In particular, the synergy around the activities of the working group STAPH is a permanent source of inspiration in these statistical functional fields.

This book presents in a original way new nonparametric statistical methods for functional data analysis. Because we support the idea that statistics should take advantage of interactions between applied and theoretical aspects, we deliberately decided not to privilege one over the other. So, this work proposes two levels of reading. Recent theoretical advances, as far as possible, are presented in self-contained sections while statistical methodology, practical aspects, and elementary mathematics are accessible to a very large public. But, in any case, each part of this book starts with the presentation of general ideas concerning theoretical as well as applied issues.

This book could be useful as well for practitioners as for researchers and students. Non expert researchers and students will find detailed proofs and mathematical tools for theoretical advances presented in this book. For experienced researchers, these advances have been selected to balance the trade-off between comprehensive reading and up-to-date results. Because nonparametric functional statistics is a recent field of research, we discuss the existing bibliography by emphasizing open problems. This could be the starting point for further statistical developments. Practitioners will find short descriptions on how to implement the proposed methods while the companion website (*http://www.lsp.ups-tlse.fr/staph/npfda*) includes large details for codes, guidelines, and examples of use. So, the use of such nonparametric functional procedures will be easy for any users. In this way, we can say that this book is really intended for a large public: practitioners, theoreticians and anybody else who is interested in both aspects.

The novelty of nonparametric functional statistics obliges us to start by clarifying the terminology, by presenting the various statistical problems and by describing the kinds of data (mainly curves). Part I is devoted to these generalities. The remaining parts consist in describing the nonparametric statistical methods for functional data, each of them being basically split into theoretical, applied, and bibliographical issues. Part II focuses on prediction problems involving functional explanatory variables and scalar response. We study regression, conditional mode and conditional quantiles and their kernel nonparametric estimates. Part III concerns the classification of functional data. We focus successively on curve discrimination (prediction of a categorical response corresponding to the class membership) and unsupervised classification (i.e., the class membership is unobserved). Because time series can be viewed as a particular case of functional dataset, we propose in Part IV to extend most of the previous developments to dependent samples of functional data. The dependance structure will be taken into account through some mixing notion. In order to keep the main body of the text clear, theoretical tools are put at the end of this monograph in the appendix.

All the routines are implemented in the R and S+ languages and are available on the companion website (*http://www.lsp.ups-tlse.fr/staph/npfda*). S+ is an object-oriented language intensively used in engineering and applied mathematical sciences. Many universities, intitutions and firms use such a software which proposes just as well a very large number of standard statistical methods as a programming language for implementing and popularizing new ones. In addition, all subroutines are translated into R because many other people work with such software, which is a free-version of S+ developed by academic researchers.

Science finds its source in the collective knowledge which is based on exchanges, collaborations and communications. So, as with any scientific production, this book has taken many benefits from contacts we had along the last few years. We had the opportunity to collaborate with various people including A. Ait-Saidi, G. Aneiros, J. Boularan, C. Camlong, H. Cardot, V. Couallier, S. Dabo-Niang, G. Estévez, W. Gonzalez-Manteiga, L. Györfi, A. Goia, W. Härdle, J. Hart, I. Horova, R. Kassa, A. Laksaci, A. Mas, S. Montcaup, V. Nuñez-Antón, L. Pélégrina, A. Quintela del Rio, M. Rachdi, J. Rodriguez-Poo, P. Sarda, S. Sperlicht and E. Youndjé, and all of them have in some sense indirectly participated to this work. Many other statisticians including J. Antoch, D. Bosq, A. Cuevas, A. Kneip, E. Kontoghiorghes, E. Mammen, J.S. Marron, J. Ramsay and D. Tjostheim have also been useful and fruitful supports for us.

Of course, this book would not have became reality without the permanent encouragements of our colleagues in the working group STAPH in Toulouse. This group acting on functional and operatorial statistics is a source of inspiration and in this sense, A. Boudou, H. Cardot, Y. Romain, P. Sarda and S. Viguier-Pla are also indirectly involved in this monograph. We would also like to express our gratitude to the numerous participants in the activities of

STAPH, with special thanks to J. Barrientos-Marin and L. Delsol for their previous reading of this manuscript and their constructive comments.

Gérard Collomb (1950-1985) was a precursor on nonparametric statistics. His international contribution has been determinant for the development of this discipline, and this is particularly true in Toulouse. Undoubtly, his stamp is on this book and we wish to take this opportunity for honoring his memory.

Frédéric Ferraty
Philippe Vieu
Toulouse, France
January, 2006

Contents

List of Abbreviations and Symbols

$[a,b]$	closed interval of $\mathbb{R}$
(a,b)	open interval of $\mathbb{R}$
$[a,b), (a,b]$	semi-open intervals of $\mathbb{R}$
a.co.	almost complete (convergence) for sequence of r.r.v.
$B(\chi,h)$	open ball of center χ and radius h, in the space (E,d)
c.d.f.	cumulative distribution function
C or C'	generic finite real positive constants
C_G^0	set of real-valued continuous functions defined on G
$d(.;.)$	semi-metric on some functional space E
$Diam(A)$	diameter of some subset A of (E,d)
E or F	generic functional spaces
(E,d)	generic functional space and its semi-metric
$\mathbb{E}(Y)$ or $\mathbb{E}Y$	expectation of some r.r.v. Y
$\mathbb{E}(Y|\boldsymbol{\mathcal{X}}=\chi)$	conditional expectation of some r.r.v. Y given the f.r.v. $\boldsymbol{\mathcal{X}}$
f.r.v.	functional random variable
f	marginal density of the f.r.v. $\boldsymbol{\mathcal{X}}$
$F_Y^{\boldsymbol{\mathcal{X}}}(\chi,y)$ or $F_Y^\chi(y)$	conditional distribution of some r.r.v. Y given the f.r.v. $\boldsymbol{\mathcal{X}}$
$f_Y^{\boldsymbol{\mathcal{X}}}(\chi,y)$ or $f_Y^\chi(y)$	conditional density of some r.r.v. Y given the f.r.v. $\boldsymbol{\mathcal{X}}$
$\phi_\chi(h)$	measure of $B(\chi,h)$ respect to the probability law of $\boldsymbol{\mathcal{X}}$
$1_I(.)$	indicator function on some set I
h	generic notation for a bandwidth $h=h(n)$
H	generic notation for an integrated kernel function
K	generic notation for an asymetrical kernel function
K_0	generic notation for a standard symetrical kernel function
$Lip_{G,\beta}$	set of real-valued Hölder continuous functions defined on G
μ	generic notation for a measure on some infinite dimensional space
$\mathbb{N}, \mathbb{N}_*$	set of positive integers with (respectively without) 0

$O_{a.co.}$	rate of almost complete convergence
$O_{a.s.}$	rate of almost sure convergence
O_p	rate of convergence in probability
$(\Omega, \mathcal{A}, P)$	probability space on which all the r.v. are defined
$P(.|\boldsymbol{\mathcal{X}} = \chi)$	conditional (on the f.r.v. $\boldsymbol{\mathcal{X}}$) probability measure on Ω
r	nonlinear regression operator
S	generic compact subset of $\mathbb{R}$
$\mathcal{S}$	the statistical functional sample $\mathcal{S} = \{\boldsymbol{\mathcal{X}}_1, \ldots, \boldsymbol{\mathcal{X}}_n\}$
X	generic random real variable
x	generic (unrandom) real number
$X_i, \; i = 1 \ldots n$	sample of r.r.v.
$x_i, \; i = 1 \ldots n$	statistical observations of the r.r.v. X_i
$\mathbf{X}$	generic multivariate random variable
$\mathbf{x}$	generic (unrandom) multivariate vector
$\boldsymbol{X_i}, \; i = 1 \ldots n$	sample of random vectors
$\boldsymbol{x_i}, \; i = 1 \ldots n$	statistical observations of the random vectors $\boldsymbol{X_i}$
$\boldsymbol{\mathcal{X}}$	generic functional random variable
χ	generic (unrandom) functional element
$\boldsymbol{\mathcal{X}}_i, \; i = 1 \ldots n$	sample of f.r.v.
$\chi_i, \; i = 1 \ldots n$	statistical observations of the f.r.v. $\boldsymbol{\mathcal{X}}_i$
r.r.v.	real random variable
$\mathbb{R}, \mathbb{R}_*$	set of real numbers with (respectively without) 0
$\mathbb{R}^+, \mathbb{R}_*^+$	set of positive real numbers with (respectively without) 0
$r.v.$	random variable
$\mathbb{Z}, \mathbb{Z}_*$	set of integers with (respectively without) 0

List of Figures

Part I

Statistical Background for Nonparametric
Statistics and Functional Data

Our main goal is to propose a new methodology combining both non-parametric modelling and functional data. Because of the obvious theoretical difficulties, despite its practical and theoretical interests, such a blend of topics was still considered unrealistic just a few years ago. This is well summarized by two citations coming from authors well-known for their statistical contributions in particular concerning functional data (parametric) modelling. The first is due to [B91] in a paper dealing with autoregressive hilbertian processes: *"These being nonparametric by themselves, it seems rather heavy to introduce a nonparametric model for observation lying in functional space ..."*. The second one can be found in the book of [RS97] *Functional Data Analysis* in the section entitled "Challenges for the future": *"theoretical aspects of Functional Data Analysis have not been researched in sufficient depth, and it is hoped that appropriate theoretical developments will feed back into advances in practical methodology."* Our aim is to take up these challenges, by giving both theoretical and practical supports for attacking functional data analysis and modelling in a free-parameter fashion.

Because of the novelty of this field, it is necessary to start by clarifying the vocabulary linking functional and nonparametrical statistics (what are functional data/variables? what is a nonparametric model for such a dataset?...). This is done in Chapter 1. Chapter 2 presents various statistical problems together with several functional data chosen in order to cover different fields of applied statistics. Chapter 3 highlightes the usefulness of the introduction of semi-metrics for modelling functional data. This turns out to be a sufficiently general theoretical framework without being *"too heavy"* in terms of computational issues. Basic statistical ideas on local weighting methods and their extension to the functional case are exposed in Chapter 4. Special attention is paid to kernel weighting.

Finally, note that all the chapters in this part are accessible to a large public. However, Chapter 2 is rather oriented towards practitioners whereas Chapter 4 focuses on preliminary technical support. Chapter 3 gives some basic considerations which are really at the interface between applied and mathematical statistics, and therefore should interest a large public.

1

Introduction to Functional
Nonparametric Statistics

The main goal of this chapter is to familiarize the reader with both functional and nonparametric statistical notions. First and because of the novelty of this field of research, we propose some basic definitions in order to clarify the vocabulary on both functional data/variables and nonparametric modelling. Second, we fix some notations to unify the remaining of the book.

1.1 What is a Functional Variable?

There is actually an increasing number of situations coming from different fields of applied sciences (environmetrics, chemometrics, biometrics, medicine, econometrics, . . .) in which the collected data are curves. Indeed, the progress of the computing tools, both in terms of memory and computational capacities, allows us to deal with large sets of data. In particular, for a single phenomenon, we can observe a very large set of variables. For instance, look at the following usual situation where some random variable can be observed at several different times in the range (t_{min}, t_{max}). An observation can be expressed by the random family $\{X(t_j)\}_{j=1,...,J}$. In modern statistics, the grid becomes finer and finer meaning that consecutive instants are closer and closer. One way to take this into account is to consider the data as an observation of the continuous family $\mathcal{X} = \{X(t); \ t \in (t_{min}, t_{max})\}$. This is exactly the case of the speech recognition dataset that we will treat deeply later in this monograph (see Section 2.2). Of course, other situations can be viewed similarly such as for instance the spectrometric curves presented in Section 2.1, for which the measurements concern different wavelengths instead of time points. Moreover, a new literature is emerging which deals with sparse functional data. In this situation, the number of measurements is small but the data are clearly of functional nature (see for instance the electrical consumption curves described in Section 2.3). To fix the ideas, we give the following general definition of a functional variable/data.

> **Definition 1.1.** *A random variable $\mathcal{X}$ is called functional variable (f.v.) if it takes values in an infinite dimensional space (or functional space). An observation χ of $\mathcal{X}$ is called a functional data.*

Note that, when $\mathcal{X}$ (*resp.* χ) denotes a random curve (*resp.* its observation), we implicitly make the following identification $\mathcal{X} = \{\mathcal{X}(t); \ t \in T\}$ (*resp.* $\chi = \{\chi(t); \ t \in T\}$). In this situation, the functional feature comes directly from the observations. The situation when the variable is a curve is associated with an unidimensional set $T \subset \mathbb{R}$. Here, it is important to remark that the notion of functional variable covers a larger area than curves analysis. In particular, a functional variable can be a random surface, like for instance the grey levels of an image or a vector of curves (and in these cases T is a bidimensional set $T \subset \mathbb{R}^2$), or any other more complicated infinite dimensional mathematical object. Even if the real data used as supports throughout this book are all curves datasets (i.e., a set of curves data), all the methodology and theoretical advances to be presented later are potentially applicable to any other kind of functional data.

1.2 What are Functional Datasets?

Since the middle of the nineties, the increasing number of situations when functional variables can be observed has motivated different statistical developments, that we could quickly name as *Statistics for Functional Variables* (or *Data*). We are determinedly part of this statistical area since we will propose several methods involving statistical functional sample $\mathcal{X}_1, \ldots, \mathcal{X}_n$. Let us start with a precise definition of a functional dataset.

> **Definition 1.2.** *A functional dataset $\chi_1, \ldots, \chi_n$ is the observation of n functional variables $\mathcal{X}_1, \ldots, \mathcal{X}_n$ identically distributed as $\mathcal{X}$.*

This definition covers many situations, the most popular being curves datasets. We will not investigate the question of how these functional data have been collected, which is linked with the discretization problems. According to the kind of the data, a preliminary stage consists in presenting them in a way which is well adapted to functional processing. As we will see, if the grid of the measurements is fine enough, this first important stage involves usual numerical approximation techniques (see for instance the case of spectrometric data presented in Chapter 3). In other standard cases, classical smoothing methods can be invoked (see for instance the phonemes data and the electrical consumption curves discussed in Chapter 3). There exist some other situations which need more sophisticated smoothing techniques, for instance when the repeated measures per subjects are very few (sparse data) and/or with irregular grid. This is obviously a parallel and complementary field of research but this is far from our main purpose which is nonparametric statistical treatments of functional data. From now on, we will assume that we have at hand a sample of functional data.

1.3 What are Nonparametric Statistics for Functional Data?

Traditional statistical methods fail as soon as we deal with functional data. Indeed, if for instance we consider a sample of finely discretized curves, two crucial statistical problems appear. The first comes from the ratio between the size of the sample and the number of variables (each real variable corresponding to one discretized point). The second, is due to the existence of strong correlations between the variables and becomes an ill-conditioned problem in the context of multivariate linear model. So, there is a real necessity to develop statistical methods/models in order to take into account the functional structure of this kind of data. Most of existing statistical methods dealing with functional data use linear modelling for the object to be estimated. Key references on methodological aspects are those by [RS97] and [RS05], while applied issues are discussed by [RS02] and implementations are provided by [CFGR05]. Note also that, for some more specific problem, some theoretical support can be found in [B00].

On the other hand, nonparametric statistics have been developped intensively. Indeed, since the beginning of the sixties, a lot of attention has been paid to free-modelling (both in a free-distribution and in a free-parameter meaning) statistical models and/or methods. The functional feature of these methods comes from the nature of the object to be estimated (such as for instance a density function, a regression function, ...) which is not assumed to be parametrizable by a finite number of real quantities. In this setting, one is usually speaking of **Nonparametric Statistics** for which there is an abundant literature. For instance, the reader will find in [H90] a previous monograph for applied nonparametric regression, while [S00] and [AP03] present the state of the art in these fields. It appears clearly that these techniques concern only classical framework, namely real or multidimensional data.

However, recent advances are mixing nonparametric free-modelling ideas with functional data throughout a double infinite dimensional framework (see [FV03b] for bibliography). The main aim of this book is to describe both theoretical and practical issues of these recent methods through various statistical problems involving prediction, time series and classification. Before to go on, and in order to clarify the sense of our purpose, it is necessary to state precisely the meanings of the expressions parametric and nonparametric models.

There are many (different) ways for defining what is a nonparametric statistical model in finite dimensional context, and the border between nonparametric and parametric models may sometimes appear to be unclear (see the introduction in [BL87] for more discussion). Here, we decided to start from the following definition of nonparametric model in finite dimensional context.

Definition 1.3. *Let X be a random vector valued in $\mathbb{R}^p$ and let ϕ be a function defined on $\mathbb{R}^p$ and depending on the distribution of X. A model for the estimation of ϕ consists in introducing some constraint of the form*

$$\phi \in \mathcal{C}.$$

The model is called a parametric model for the estimation of ϕ if $\mathcal{C}$ is indexed by a finite number of elements of $\mathbb{R}$. Otherwise, the model is called a nonparametric model.

Our decision for choosing this definition was motivated by the fact that it makes definitively clear the border between parametric and nonparametric models, and also because this definition can be easily extended to the functional framework.

Definition 1.4. *Let $\mathcal{Z}$ be a random variable valued in some infinite dimensional space F and let ϕ be a mapping defined on F and depending on the distribution of $\mathcal{Z}$. A model for the estimation of ϕ consists in introducing some constraint of the form*

$$\phi \in \mathcal{C}.$$

The model is called a functional parametric model for the estimation of ϕ if $\mathcal{C}$ is indexed by a finite number of elements of F. Otherwise, the model is called a functional nonparametric model.

The appellation **Functional Nonparametric Statistics** covers all statistical backgrounds involving a nonparametric functional model. In the terminology **Functional Nonparametric Statistics**, the adjective **nonparametric** refers to the form of the set of constraints whereas the word **functional** is linked with the nature of the data. In other words, nonparametric aspects come from the infinite dimensional feature of the object to be estimated and functional designation is due to the infinite dimensional feature of the data. That is the reason why we may identify this framework to a double infinite dimensional context. Indeed, ϕ can be viewed as a nonlinear operator and one could use the terminology *model for operatorial estimation* by analogy with the multivariate terminology *model for functional estimation*.

To illustrate our purpose concerning these modelling aspects, we focus on the regression models

$$Y = r(X) + error, \tag{1.1}$$

where Y is a real random variable by considering various situations: linear (parametric) or nonparametric regression models with curves (i.e. $X = \mathcal{X} = \{X(t); \ t \in (0,1)\}$) or multivariate (i.e. $X = \boldsymbol{X} = (X^1, \ldots, X^p)$) data:

		MODELS	
		LINEAR	NONPARAMETRIC
D	MULTIVARIATE	*Example 1* $X \in \mathbb{R}^p$ $\mathcal{C} = \{r \text{ linear}\}$	*Example 2* $X \in \mathbb{R}^p$ $\mathcal{C} = \{r \text{ continuous}\}$
A T A	FUNCTIONAL	*Example 3* $X \in F = L^2_{(0,1)}$ $\mathcal{C} = \{\chi \mapsto \int_0^1 \rho(t)\chi(t)dt \in \mathbb{R}, \ \rho \in F\}$	*Example 4* $X \in F = L^2_{(0,1)}$ $\mathcal{C} = \{r \text{ continuous}\}$

Example 1 corresponds to the so-called multivariate linear regression model

$$Y = a_0 + \sum_{j=1}^{p} a_j\, X^j + error,$$

which is obviously a parametric model (with $p + 1$ unknown real parameters $a_0, \ldots, a_p$). Example 2 refers to the classical multivariate nonparametric regression model

$$Y = r(X^1, \ldots, X^p) + error.$$

Now, Example 3 is exactly what [RS97] call *functional linear regression model for scalar response* namely

$$Y = \int_0^1 \rho(t)X(t)dt + error$$

which can be reformulated as (1.1) with r being a continuous linear operator from F to $\mathbb{R}$ (by using the Riesz representation theorem). According to our definition, this is a functional parametric regression model, where $\rho(.)$ is the only one (functional) parameter. The last model (Example 4) can be written as (1.1) where r is just a continuous operator from F to $\mathbb{R}$. Example 4 is a functional nonparametric regression model according to Definition 1.4. This model will be treated with details in Chapter 5.

1.4 Some Notation

In the remaining of this book, χ will denote any non-random element of some infinite dimensional space E and $\boldsymbol{\mathcal{X}}$ a functional random variable valued in E. Similarly, $\{\chi_i\}_{i=1,\ldots,n}$ will denote the n observations of a sample $\{\boldsymbol{\mathcal{X}}_i\}_{i=1,\ldots,n}$ of f.r.v. Even if this monograph is devoted to the study of the nonparametric method for functional data, we will still need to introduce real or multivariate variables. Instead of χ, $\boldsymbol{X}$, χ_i and $\boldsymbol{X}_i$, we will use boldfaced letters for vectors ($\boldsymbol{x}$, $\boldsymbol{X}$, $\boldsymbol{x_i}$ and $\boldsymbol{X_i}$) and standard letters for the real case (x, X, x_i, X_i).

In addition, any random variable considered in this book is defined on the same probability space $(\Omega, \mathcal{A}, P)$. Finally, except in Part IV, any statistical sample is implicitly assumed to be independent.

1.5 Scope of the Book

Once the general framework for nonparametric modelling of functional variable is given (see Part I), the book focuses on various statistical topics: predicting from a functional variable (see Part II), classifying a sample of functional data (see Part III) and statistics for dependent functional variables (see Part IV). All these statistical methods are developed in a free-parameter way (which includes free-distribution modelling). In this sense, this book is both completely different and complementary to the few other books existing on functional data ([RS97], [RS02] and [RS05]). Rather than set application against theory, this book is really an interface of these two features of statistics and each statistical topic is investigated from its deep theoretic foundations up to computational issues. For more details on practical aspects, the reader can refer to the companion website *http://www.lsp.ups-tlse.fr/staph/npfda*). This site contains functional datasets, case studies and routines written in two popular statistical languages: R (see [RDCT]) and S+ (see the comprehensive manual of [BCW88], as well as more recent literature in [C98], [KO05] or [VR00]).

2

Some Functional Datasets and Associated Statistical Problematics

This chapter could have been entitled "Some statistical problematics and associated functional data". In fact, there are many nonparametrical statistical problems which occur in the functional setting. Sometimes, they appear purely in terms of statistical modelling or, on the contrary, they can be drawn directly from some specific functional datasets. But, in any case, the proposed solutions should look at both points of view. This chapter describes various functional data with their associated statistical problems. Of course, some statistical processing (such for instance unsupervised classification) concerns all the following datasets but additional informative data (such for instance the knowledge of some response variable) can lead to particular problems. As we will see, these data have been choosen to cover different applied statistics fields, different shapes of curves (smooth, unsmooth), various grids of discretization (fine, sparse) and also different types of statistical problems (regression, discrimination, prediction and classification). Obviously, the methods presented later on will concern many other functional datasets while at the same time, these datasets can motivate other statistical problems not investigated here. Although these functional datasets are available on various websites, we give them in the companion website (*http://www.lsp.ups-tlse.fr/staph/npfda*) in which they are presented in a appropriate format, directly usable by readers both for familiarizing themselves (by reproducing the examples) with the functional nonparametric methods described in this monograph and eventually to compare them with their own alternative approaches.

2.1 Functional Chemometric Data

Spectrometry is a modern and useful tool for analyzing the chemical composition of any substance. As pointed out by [FF93], in order to analyze such kind of data, "chemometricians have invented their own techniques based on heuristic reasoning and intuitive ideas". The two most popular methods are partial least squares (see [W75] and [MNa89] for more details), and principal component regression ([M65]). As chemometrics was a starting point for developing the functional nonparametric methodology, they play a major role in this book, and it is natural to begin by the presentation of such dataset.

2.1.1 Description of Spectrometric Data

The original data come from a quality control problem in the food industry and can be found at *http://lib.stat.cmu.edu/datasets/tecator*. Note that they were first studied by [BT92] using a neural networks approach. This dataset concerns a sample of finely chopped meat. Figure 2.1 displays some units among the original spectrometric data.

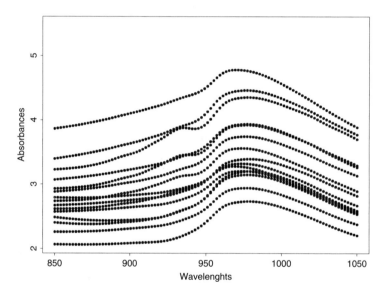

Fig. 2.1. Original Chemometric Data Concerning 15 Subjects

This figure plots absorbance versus wavelength for 15 randomly selected pieces of meat. More precisely, for each meat sample the data consists of a 100 channel spectrum of absorbances. Absorbance is the $-log_{10}$ of the transmittance measured by the spectrometer, and the data were recorded on a Tecator Infractec Food and Feed Analyzer working in the wavelength range 850-1050 nm by the near-infrared (NIR) transmission principle. One unit appears clearly as a discretized curve. Because of the fineness of the grid (spaning the discretization), we can consider each subject as a continuous curve. This was pointed out by [LMS93] who said of such chemometric data that "the spectra observed are to all intents and purposes functional observations". Thus, each spectrometric analysis can be summarized by some continuous curve giving the observed absorbance as function of the wavelength. The curves dataset, that is the whole set of (continuous) data, is presented in Figure 2.2 below.

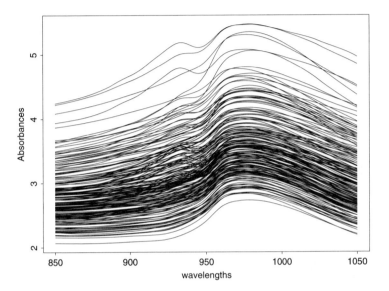

Fig. 2.2. The Spectrometric Curves

As we can see in Figure 2.2, the shape of these spectrometric curves is very smooth and looks very similar except for an obvious vertical shift.

2.1.2 First Study and Statistical Problems

Considering these data, many questions can appear. In particular, the first natural one would be to know whether the vertical shift observed in Figure 2.2 is really informative or not. In other words, does the shift come from the chemical components of the meat or is it only due to some extra phenomenon? This question is fundamental because the shift can hide other important features and act as a trompe-l'oeil. We will see in Chapter 3 that the functional modelization of these data allows us to take care of this shift effect and finally to furnish some arguments to answer these questions. But let us first explore the data by means of classical Principal Component Analysis (PCA) which is a very popular and useful tool in terms of multivariate factor analysis. Figures 2.3 displays both the 100 variables and the 215 units on the principal eigenspace.

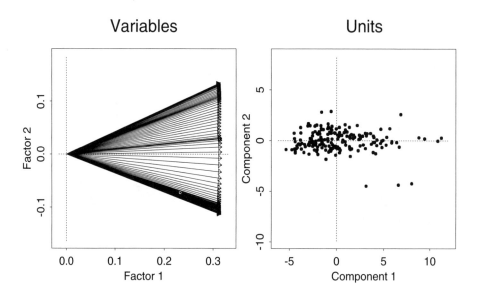

Fig. 2.3. Standard PCA: Spectrometric Data

From the variables displayed, it appears that there is a clear scale factor, and this could be probably be linked with the vertical shift. Concerning the individuals, we just note some potential outliers but no strong structure seems to appear. Obviously, pertinent information can be found by looking at the remaining dimensions (axes 3, 4,..., 100). But here, we see some limits of this traditional statistical technique due to the high dimensionality of the data.

After this explanatory stage, and since the role of the shift effect seems to be identified, many statistical questions may arise. For instance, one could wish to know whether the curves are similar or not. This is an important problem because the sample of curves may hide several different subsamples. This is typically an unsupervised curve classification problem, for which we will see the usefulness of a functional nonparametric approach as detailed in Chapter 9. Another question comes from the fact that the main goal of spectrometric analysis is to allow for the discovery of the proportion of some specific chemical content whereas the analytic chemistry processing would take more time and be much more expensive. For instance, if Y is the proportion of some component (in our dataset, Y is the percentage of fat content in the piece of meat), one would like to use the spectrometric curves to predict Y. This is typically a functional regression problem with scalar response and Chapter 5 will show how the nonparametric approach provides some nice tools for that.

2.2 Speech Recognition Data

2.2.1 What are Speech Recognition Data?

In speech recognition, the observed data are clearly of a functional nature. For instance, look at the following data set, previously introduced by a joint collaboration between Andreas Buja, Werner Stuetzle and Martin Maechler, and used as illustration by [HBT95] (data and description are available at *http:www-stat.stanford.edu/ElemStatLearn*, which is the website of [HBF01]). The data are log-periodograms corresponding to recordings of speakers of 32 ms duration. Here also, even if we have to deal with discretized data, the number of observed points is quite large enough to allow for considering the observations to be continuous (as they are, indeed). The study concerns (see Figure 2.4) five speech frames corresponding to five phonemes transcribed as follows:

- "sh" as in "she" (group 1);
- "iy" as in "she" (group 2);
- "dcl" as in "dark" (group 3);
- "aa" as the vowel in "dark" (group 4);
- "ao" as the first vowel in "water" (group 5).

Precisely, our dataset is extracted from the original one in such a way that each speech frame is represented by 400 samples at a 16-kHz sampling rate; only the first 150 frequencies from each subject are retained, and the data consist of 2000 log-periodograms of length 150, with known class phoneme memberships. Indeed, Figure 2.4 displays only 10 log-periodogram curves for each class phoneme.

2.2.2 First Study and Problematics

We can start the study of these data with an exploratory stage. Once again, we use the classical PCA in order to discover some structure. Figure 2.5 displays

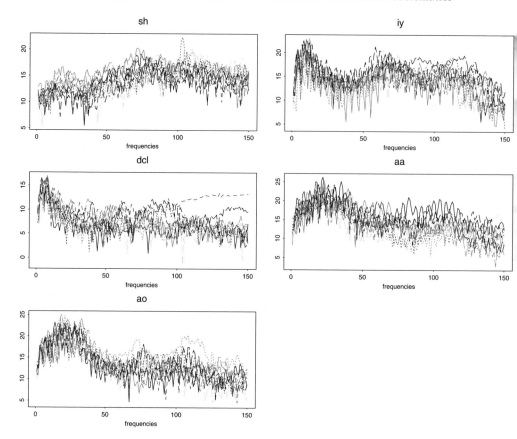

Fig. 2.4. A Sample of 10 Log-Periodograms (Curves Data) for each of the Five Phoneme Classes

both the 150 variables and the 2000 units on different eigenspaces. Following the results of this factorial analysis, eigenspaces $(1, 2)$ and $(1, 3)$ exhibit just one group which corresponds to the sound "dcl". One can remark too that axis 2 seems to draw a "fuzzy" border between the groups "sh/iy" and "aa/ao".

After this first analysis, one can pursue this exploratory step in order to extract more pertinent information. Some interesting questions are: is there a common structure? Can we split the log-periodograms into several classes? One way to answer is to consider an unsupervised classification problem. But here, we have important additional information since we have the knowledge of the classes for the sample of log-periodograms. Thus, we can give answers to the previous questions by solving a curve discrimination problem or a supervised curve classification problem. That is the aim of Chapter 8. As we will see, nonparametric methods are well adapted to this kind of situation because we have no idea about the relationship between log-periodograms and sounds.

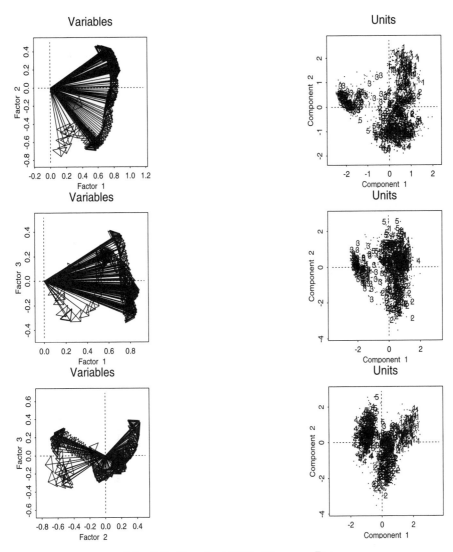

Fig. 2.5. Standard PCA: Phoneme Data

2.3 Electricity Consumption Data

2.3.1 The Data

This section focuses on an economic time series: the US monthly electricity consumed by the residential and commercial sectors from January 1973 to February 2001 (338 months). The data are available on the web site *http://www.economagic.com*. Figure 2.6 displays this time series which obviously exhibits some linear trend, as well as some heterogeneity in the variance structure.

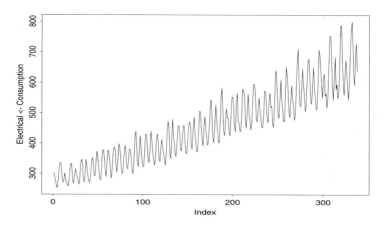

Fig. 2.6. The Electricity Consumption Data

In order to clean these effects, we eliminated the heteroscedasticity and the linear trend by differenciating the log-data. Figure 2.7 displays the transformed time series.

2.3.2 The Forecasting Problematic

One of the main specific problems in these situations is to predict future consumption, and usual statistical models (either parametric or nonparametric) achieve that by taking into consideration a finite number of past data. However, one could think that it is more reasonable to take into account as explanatory variable the continuous time series over some period (see Chapter 12 for more details). For our example, we decided to choose the whole past year as explanatory period. That means that the set of explanatory variables to be included in our statistical method is composed with 28 curves data which are the 28 yearly continuous time series. These functional data are presented in Figure 2.8.

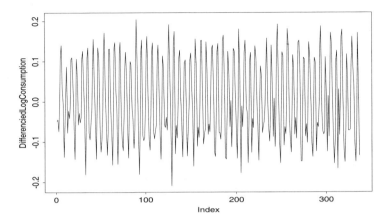

Fig. 2.7. Electricity Consumption: Differenciated Log Data

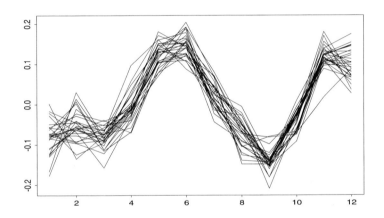

Fig. 2.8. Electricity Consumption: Yearly Differenciated Log Curves

A first study consists in achieving a factorial analysis on these 28 curves. Figure 2.9 displays the first eigenspace both for variables and units. It is clear here that there is no visible structure. This fact confirms that the past of our time series cannot be summarized by a small number of parameters. In other words, the whole past year should be incorporated for predicting future values.

If we look at this as matrix data coming from 12 variables and 28 subjects, there is an obvious problem due to low ratio units/variables. So, even if the measurements per period are few in comparison with the previous spectro-

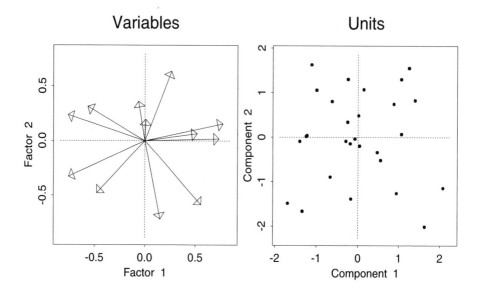

Fig. 2.9. Standard PCA: Electricity Consumption Data

metric and phonetic datasets, we will see in Chapter 11 that functional non-parametric approaches work well.

Of course, there are many other interesting problems to solve with this kind of data. For instance, we can wish to classify the 28 trajectories in an unsupervised way in order to answer the question: can we split the year into several groups? If, in addition, we take into account a categorical variable (for instance, the supremum of a path is greater than a fixed threshold), the classification aim becomes a curve discrimination problem.

3

What is a Well-Adapted Space for Functional Data?

Using functional data asks crucial statistical questions. Indeed, the larger is the space E in which the variable takes its values, the sparser are the data. In the case of functional data, we know (by the nature itself of the data) that E is an infinite dimensional space. So, this chapter focuses on this essential problem of high (i.e., infinite) dimensional data. Obviously, the sparseness notion is strongly linked with the way used to measure closeness between data, and we propose an original way to approach this question by mean of semi-metric considerations.

3.1 Closeness Notions

Proximities measures between mathematical objects play a major role in all statistical methods. In many situations, a classical norm can be used to measure the closeness between two objects. Because in a finite dimensional euclidean space (typically $\mathbb{R}^p$) there is an equivalence between all norms, the choice in the mathematical sense of this kind of measure is not crucial apart from some practical constraints (as, for instance, computational ease). Once a preliminary norm is fixed, it is clear that we can deduce a family of norms and from a statistical point of view, there remains one essential question: namely the choice among these different metrics. For instance, one of the most popular in $\mathbb{R}^p$ is the usual euclidean norm $\|.\|$ which is based on the sum of squares of the components of any vector. More precisely, let $\boldsymbol{x} = {}^t(x_1, \ldots, x_p)$ be a vector of $\mathbb{R}^p$; then, the classical euclidean norm is defined by

$$\|\boldsymbol{x}\|^2 = \sum_{j=1}^{p}(x_j)^2 = {}^t\boldsymbol{x}\,\boldsymbol{x}.$$

Of course, we can deduce a family of norms based on the euclidean norm by using different definite positive matrices $\boldsymbol{M}$, in the following way

$$\|x\|_M^2 = {}^t x \, M \, x.$$

The choice of the norm comes to the same as the choice of M.

Now, considering an infinite dimensional space, the equivalence between norms fails and the problem has to be attacked in a different way. In other words, in the functional context, the choice of the preliminary norm becomes crucial. Even more, considering normed or metric spaces can become too restrictive. In some situations and this is the case for our datasets, it appears that semi-metric spaces are better adapted than metric spaces. As we will see later, the shape of data and eventually exogene informations or the goal of the statistical study can help to drive the semi-metric selection. The aim of the next section is to show the benefit of considering semi-metrics as a closeness measure. Before going on, let us just recall some basic definitions.

Definition 3.1. $\|.\|$ *is a semi-norm on some space F as soon as:*
1) $\forall (\lambda, x) \in \mathbb{R} \times F, \ \|\lambda x\| = |\lambda| \, \|x\|$
2) $\forall (x, y) \in F \times F, \ \|x + y\| \leq \|x\| + \|y\|.$

Note that in fact, a semi-norm $\|.\|$ is a norm except that $\|x\| = 0 \not\Rightarrow x = 0$. Similarly, a semi-metric d can be defined to be a metric but such that $d(x, y) = 0 \not\Rightarrow x = y$.

Definition 3.2. d *is a semi-metric on some space F as soon as:*
1) $\forall x \in F, \ d(x, x) = 0,$
2) $\forall (x, y, z) \in F \times F \times F, \ d(x, y) \leq d(x, z) + d(z, y).$

3.2 Semi-Metrics as Explanatory Tool

A large part of explanatory tools consists in displaying data in low-dimensional spaces. It is clear that the shape of such graphics depends strongly on the proximity measure. Look at the chemometric dataset. As was shown in Section 2.1, one can start the study by usual Principal Component Analysis (PCA), the proximity between subjects (spectrometric curves) being computed by means of the classical L_2-metric, which is defined for all observed curves χ_i and $\chi_{i'}$ by

$$\sqrt{\left(\int \left(\chi_i(t) - \chi_{i'}(t) \right)^2 \, dt \right)}.$$

Because the first axis has been interpreted as a factor scale (see Figure 2.3), it is pertinent to see the eigenspace spanned by axes 2 and 3 (see Figure 3.1).

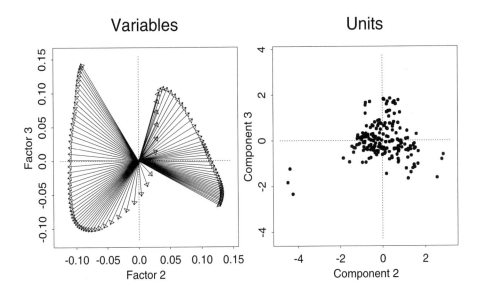

Fig. 3.1. PCA with L_2-metric; Axes 2 and 3

However, one can propose other measures of proximity. For instance, consider the following closeness measure based on the second derivative:

$$\sqrt{\int \left(\chi_i^{(2)}(t) - \chi_{i'}^{(2)}(t)\right)^2 dt}.$$

Note that this is not a metric but only a semi-metric according to Definition 3.2. Now, we can achieve again a Principal Component Analysis by using this semi-metric as a measure of proximity (see Figure 3.2). It is interesting to compare this graphical display with the previous one coming from the Principal Component Analysis using the L_2-metric (see Figure 2.3). There are large differences: the semi-metric approach allows us to see much more structure in the variables. Even if we display an additional axis (see Figure 3.1), the L_2-metric PCA does not reveal strong structure in the variable.

Clearly, mathematical objects for computing proximities between curves play a major role. So the question is: how to decide what is the best graphical display concerning spectrometric curves? Or equivalently: which analysis reveals the more pertinent information? One way to give a reasonable answer is

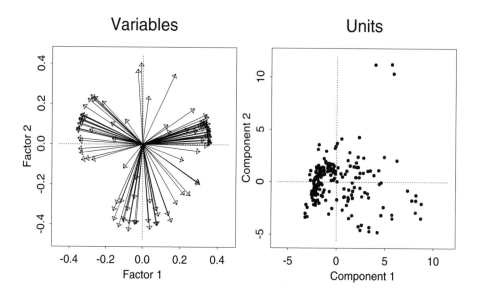

Fig. 3.2. PCA Based on Semi-Metric; Axes 1 and 2

to allow a choice inside a large family of semi-metrics. A family will be built according to each specified statistical problem and dataset. For instance, in the context of chemometric data we will focus on the family of semi-metrics.

$$\sqrt{\int \left(\chi_i^{(m)}(t) - \chi_{i'}^{(m)}(t) \right)^2 \, dt}, \ m = 0, 1, 2, \ldots, \tag{3.1}$$

where, for any m-times differentiable real function χ, $\chi^{(m)}$ denotes the mth derivative of χ (with $\chi^{(0)} = \chi$). To give an idea of the interest of this approach for the spectrometric curves, in Figure 3.3 we propose to display their successive derivatives. The main effect produced by the differentiating operator is to highlight some ranges of wavelengths with large variations. Finally, one can say that the semi-metrics act as a filter and a "good semi-metric" will be *a priori* the one which can select all the pertinent information. For the spectrometric data, we will discuss later in Section 7.2 how we can get an "optimal" semi-metric inside of the above-mentioned family.

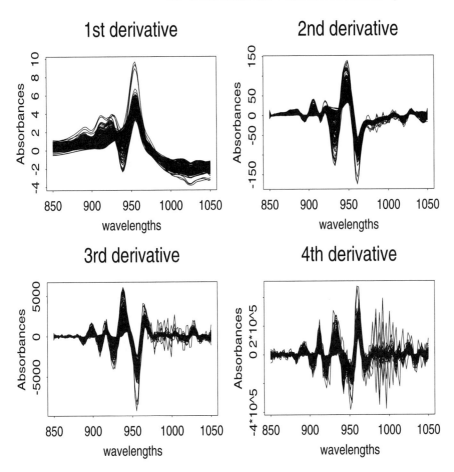

Fig. 3.3. Shape of the Derivatives of the Spectrometric Curves

3.3 What about the Curse of Dimensionality?

The curse of dimensionality is a well-known concept for nonparametricians.
This notion is strongly linked with the sparseness of data in a high-dimensional
space. An interesting question is: what about the curse of dimensionality when
we work with functional data? If we have n observations lying into $\mathbb{R}^p$, one
way to illustrate the curse of dimensionality is to count the number $N(p)$ of
units falling into a subset (of fixed size) of $\mathbb{R}^p$ when p takes successive values
$(1, 2, \ldots)$. In particular, this situation corresponds to functional data viewed
through their discretized version. Following the same idea, if we have n functional observations lying into a semi-metric space (E, d), we will count the
number N_d of units falling into a subset (of fixed size) of E.

Let $\{\chi_i = \{\chi_i(t); \ t \in (t_{min}, t_{max})\}\}_{i=1,\ldots,n}$ be a functional dataset and consider its discretized version $\{\boldsymbol{x}_i = {}^t(\chi_i(t_1), \chi_i(t_2), \ldots, \chi_i(t_J))\}_{i=1,\ldots,n}$ which can be viewed as a classical data matrix. The following array

$\chi_1(t_1)$	$\chi_1(t_2)$	$\ldots$	$\chi_1(t_J)$
$\chi_2(t_1)$	$\chi_2(t_2)$	$\ldots$	$\chi_2(t_J)$
$\vdots$	$\vdots$	$\vdots$	$\vdots$
$\chi_n(t_1)$	$\chi_2(t_2)$	$\ldots$	$\chi_n(t_J)$

can be viewed as J measurements at $t_1, \ldots, t_J$ of n observed curves (for instance, for the spectrometric curves, $J = 100$, $n = 215$, $t_{min} = 850$, $t_{max} = 1050$ and for the speech recognition data, $J = 150$, $n = 2000$, $t_{min} = 0$, $t_{max} = 150$). From this dataset, let us extract an equispaced subsequence $t_{j_1}, \ldots, t_{j_p}$ of the discretized points $t_1, \ldots, t_J$ and consider only the data corresponding to this subsequence:

$\chi_1(t_{j_1})$	$\ldots$	$\chi_1(t_{j_p})$
$\chi_2(t_{j_1})$	$\ldots$	$\chi_2(t_{j_p})$
$\vdots$	$\vdots$	$\vdots$
$\chi_n(t_{j_1})$	$\ldots$	$\chi_n(t_{j_p})$

Now, we can compute for $i = 1, \ldots, n$ the following euclidean metric

$$\delta_p(\boldsymbol{x}_i, 0) = \sum_{k=1}^{p} (\chi_i(t_{j_k}))^2$$

and let $N(p)$ be defined as

$$N(p) = \sum_{i=1}^{n} 1_{\left\{ \frac{\delta_p(\boldsymbol{x}_i, 0)}{\max_i \delta_p(\boldsymbol{x}_i, 0)} < 0.1 \right\}}.$$

Figure 3.4 displays the quantities $N(p)$ versus p for the centered spectrometric data. In addition, we replace each column of the subdataset by p independent and identically distributed standard gaussian variables and we also compute the corresponding quantities $N(p)$. The curse of dimensionality appears clearly for the datasets built with i.i.d. gaussians (the data are sparse for dimension higher than three and the sparseness increases exponentially with p) whereas the spectrometric data seem not affected by the dimension. In fact, the spectrometric data have a strong covariance structure. More precisely, because the data have been centered, a good correlation index is given by the average over i, j, and k of the quantities $\chi_i(t_j)\chi_i(t_k)$, which equals to 0.986. This strong correlation is linked with the very smooth shape of the spectrometric curves. In other words, these data can be viewed as a one-dimensional data! Similarly, Figure 3.5 produces the same plot for the phoneme dataset. Contrary to the spectrometric case, the average of the correlations equals to 0.259 which is not surprising with regard to the roughness of the Phoneme curves. So, these data

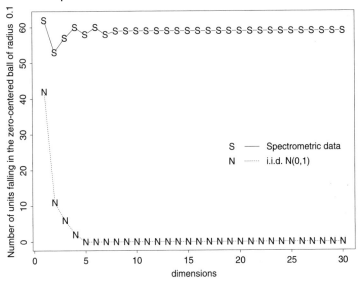

Fig. 3.4. $N(p)$ Versus p

are strongly affected by the curse of dimensionality; their sparseness looks like the one of i.i.d. variables!

Now, if we take into account the functional feature of the data, namely if we replace $x_i = (\chi_i(t_1), \ldots, \chi_i(t_J))$ by $\chi_i = \{\chi_i(t), t \in (t_{min}, t_{max})\}$, we can compute the quantity N_d defined as

$$N_d = \sum_{i=1}^{n} 1_{\left\{ \frac{d(\chi_i, 0)}{\max_i d(\chi_i, 0)} < 0.1 \right\}},$$

where d denotes a functional measure of closeness. For instance, if we consider the spectrometric curves with $d = d_2^{deriv}$ as defined in Section 3.4, we obtain $N_d = 24$ and if we look at the phoneme data with $d = d_2^{PCA}$ (as defined in Section 3.4), we get $N_d = 16$. Comparing with Figure 3.5, it appears that such a functional approach (pending of course to a right choice of the d) may override the curse of dimensionality that was observed for phoneme data.

Finally, it appears that the curse of dimensionality does not affect functional data with high correlation like the spectrometric data but is dramatic for the uncorrelated ones like the phoneme data. Nevertheless, by considering functional features, even if the data are not correlated, we partially cancel the curse of dimensionality. Of course, a crucial challenge is pointed out here: the choice of the measure of closeness d. This depends obviously on practical

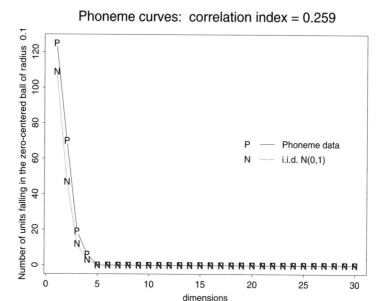

Fig. 3.5. $N(p)$ Versus p

considerations (and not only on this sparseness effect) which will be discussed later, according both to the specific statistical problem and to the set of curves.

3.4 Semi-Metrics in Practice

Because most available functional datasets are curves, we describe here semi-metrics well adapted for this kind of data. Consider a sample of curves $\mathcal{X}_1, \ldots, \mathcal{X}_n$ identically distributed as the functional random variable (f.r.v.) $\mathcal{X} = \{\mathcal{X}(t); t \in T\}$. The user has to choose among different kinds of semi-metrics which can be drawn by the shape of the curve (for instance, smooth curves could have to be treated with semi-metrics other than rough ones). We present here three families of semi-metrics but of course, many others can be built. The first two are well adapted for rough curves whereas the third one concerns quite smooth data.

3.4.1 Functional PCA: a Tool to Build Semi-Metrics

In many multivariate situations, the classical Principal Components Analysis (PCA) is considered as a useful tool for displaying data in a reduced dimensional space. More recently, the PCA methods were extended to functional data and used for many different statistical purposes. It is out of the scope of this book to make an exhaustive survey of functional PCA (FPCA) and the reader can refer, for instance, to [O60], [DPR82], [CLS86], [S96], [AOV99], [LMSTZC99], [BF00], and [EAV05]. Here, we will see that the FPCA is also a good tool for computing proximities between curves in a reduced dimensional space. As long as $E \int \mathcal{X}^2(t)dt$ is finite, the FPCA ([DPR82]) of the f.r.v. $\mathcal{X}$ allows us to obtain the following expansion of $\mathcal{X}$:

$$\mathcal{X} = \sum_{k=1}^{\infty} \left(\int \mathcal{X}(t)v_k(t)dt \right) v_k,$$

$v_1, v_2, \ldots$, being the orthonormal eigenfunctions of the covariance operator

$$\Gamma_{\mathcal{X}}(s,t) = E\left(\mathcal{X}(s)\mathcal{X}(t) \right)$$

associated with the eigenvalues $\lambda_1 \geq \lambda_2 \geq \ldots$. Now, let

$$\tilde{\mathcal{X}}^{(q)} = \sum_{k=1}^{q} \left(\int \mathcal{X}(t)v_k(t)dt \right) v_k$$

be a truncated version of the previous expansion of $\mathcal{X}$. The main interest of such a decomposition is that this truncated version is minimizing $E\left(\int (\mathcal{X}(t) - P_q\mathcal{X}(t))^2 dt \right)$ over all projections P_q of $\mathcal{X}$ into q-dimensional spaces. Thus, we can define a parametrized class of semi-norms from the classical L^2-norm in the following way:

$$||\chi||_q^{PCA} = \sqrt{\int \left(\tilde{\chi}^{(q)}(t) \right)^2 dt} = \sqrt{\sum_{k=1}^{q} \left(\int \chi(t)v_k(t)dt \right)^2}.$$

This leads to the following parametrized family of semi-metrics:

$$d_q^{PCA}(\mathcal{X}_i, \chi) = \sqrt{\sum_{k=1}^{q} \left(\int [\mathcal{X}_i(t) - \chi(t)]v_k(t)dt \right)^2}$$

Here, q is not really a smoothing parameter but rather a tuning parameter indicating the resolution level at which the problem is considered. Note that in practice, $\Gamma_{\mathcal{X}}$ is unknown and then, the $v_k's$ too, but the covariance operator can be well approximated by its empirical version

$$\Gamma_{\mathcal{X}}^n(s,t) = 1/n \sum_{i=1}^{n} \mathcal{X}_i(s)\mathcal{X}_i(t),$$

and the eigenfunctions of $\Gamma_{\mathcal{X}}^n$ are consistent estimators of those of $\Gamma_{\mathcal{X}}$ (see [CFS99]). Indeed we never observe exactly $\{\chi_i = \{\chi_i(t); t \in T\}\}_{i=1,\ldots,n}$ but only a discretized version $\{x_i = {}^t(\chi_i(t_1), \ldots, \chi_i(t_J))\}_{i=1,\ldots,n}$ (note that this is implicitly assuming that the data are balanced, which means that all units are measured at the same points). So, from a practical point of view, according to [CLS86], we can approximate the integral in the following way

$$\int [\boldsymbol{X}_i(t) - \chi(t)] v_k(t) dt \approx \sum_{j=1}^{J} w_j (\boldsymbol{X}_i(t_j) - \chi(t_j)) v_k(t_j),$$

where $w_1, \ldots, w_J$ are quadrature weights which define the approximate integration. To fix ideas, note that a standard choice could be $w_j = t_j - t_{j-1}$. If we have two discretized curves $\boldsymbol{x}_i$ and $\boldsymbol{x}_{i'}$, the quantity $d_q^{PCA}(\chi_i, \chi_{i'})$ will be approximated by its empirical version :

$$d_q^{PCA}(\boldsymbol{x}_i, \boldsymbol{x}_{i'}) = \sqrt{\sum_{k=1}^{q} \left(\sum_{j=1}^{J} w_j (\chi_i(t_j) - \chi_{i'}(t_j)) [\boldsymbol{v}_k]_j \right)^2},$$

where $\boldsymbol{v}_1, \boldsymbol{v}_2, \ldots,$ are the $\boldsymbol{W}$-orthonormal eigenvectors of the covariance matrix $(\boldsymbol{W} = diag(w_1, \ldots, w_J))$

$$\boldsymbol{\Gamma}^n \boldsymbol{W} = 1/n \sum_{i=1}^{n} \boldsymbol{x}_i \, {}^t\boldsymbol{x}_i \, \boldsymbol{W},$$

associated with the eigenvalues $\lambda_{1,n} \geq \lambda_{2,n} \geq \ldots$. Note of course that $d_q^{PCA}(\boldsymbol{x}_i, \boldsymbol{x}_{i'})$ is close to $d_q^{PCA}(\chi_i, \chi_{i'})$ as soon as the grid $(t_1, \ldots, t_J)$ is sufficiently fine.

As a conclusion, it is important to discuss what is concerning the design points. Indeed, this semi-metric can be used only if the data are balanced (the curves are observed at the same points) and the grid of the measurements sufficiently fine (see, however, Section 3.6). This could appear as a drawback for using such kind of semi-metrics, but on the other hand they have as a main advantage that they are usable even if the curves are rough. This type of semi-metrics will be adapted to such rough and balanced datasets, as for instance the phonetic data set described above.

3.4.2 PLS: a New Way to Build Semi-Metrics

The main goal is to build a new family of semi-metrics for situations when we observe an additional response, by adapting the multivariate partial least-squares regression (MPLSR) approach. Before going on, we give a brief bibliography on PLS and we recall the main idea behind the MPLSR (but it is out of our scope here to give a deep description of the PLS methods). The MPLSR is a statistical method for regressing a multivariate response (i.e., p scalar responses) on a multivariate predictor (i.e., p predictors). Partial least-squares techniques were originally developed in economical sciences ([W66]), intensively used in the chemometrician community (see [GK86]) and became popular in image processing (see [MBHG96]). The MPLSR was developed in order to predict a multivariate response from independent variables when there is a high degree of collinearity among the predictors and/or when the number of predictors is very large in comparison with the number of observations (which is the case with discretized functional data). [FF93] presents the MPLSR method as a good alternative to the *ridge* regression ([HK70]) or the *principal components* regression ([M65]). The reader can find recent works and more details in [H90], [H88] and [PT01]. MPLSR computes a simultaneous decomposition of the set of predictors and the set of responses in such a way that the performed components maximize the covariance between both sets of variables. In particular, MPLSR provides p components, each corresponding to a response, where the computed components depend on a parameter called number of factors: the larger this parameter, the better the fitting of the data. However, taking a too large number of factors can lead to components with very large variability (i.e. the noise is partially contained in the components). We can say that the number of factors performs the trade-off between accuracy and variability. In this sense, the number of factors plays similar role than the number of dimensions retained in a Principal Component Analysis (PCA). But the main difference with PCA comes from the fact that the components performed with PCA explain only the predictors whereas in the PLS approach, the components are also relevant for the multivariate response.

As for PCA, the ideas of PLS method can be useful for different purposes involving functional data (see, for instance, [PS05], and [PS05b]). In particular, we will see how PLS allows us to build a class of semi-metrics. Let $v_1^q, \ldots, v_p^q$ be the vectors of $\mathbb{R}^J$ performed by MPLSR where q denotes the number of the factors and p the number of scalar responses. Then we define the semi-metric based on the MPLSR as follows:

$$d_q^{PLS}(x_i, x_{i'}) = \sqrt{\sum_{k=1}^{p} \left(\sum_{j=2}^{J} w_j(\chi_i(t_j) - \chi_{i'}(t_j))[v_k^q]_j \right)^2},$$

where the quadrature weights $w_1, \ldots, w_J$ are those discussed in Section 3.4.1. When we consider only one scalar response ($p = 1$), the proximity between two discretized curves is due to only one direction, which seems inadequate with regard to the complexity of functional data. However, as soon as we consider multivariate response, such a family of semi-metrics allows to obtain very good results, which is the case in the curves discrimination (i.e. supervised classification) context (see Chapter 8). As with the semi-metrics based on the PCA, note that PLS type semi-metrics can be only applied on balanced data (see however Section 3.6) with a grid of the measurements sufficiently fine and are also usable even if the curves are rough like the phonetic dataset described before.

3.4.3 Semi-metrics Based on Derivatives

Another way to build a parametrized family of semi-metrics between curves is to consider a distance between one among their derivatives. More precisely, given two observed curves χ_i and $\chi_{i'}$, we consider the following semi-metric :

$$d_q^{deriv}(\chi_i, \chi_{i'})^2 = \int \left(\chi_i^{(q)}(t) - \chi_{i'}^{(q)}(t) \right)^2 dt$$

where $\chi^{(q)}$ denotes the qth derivative of χ. Note that $d_0^{deriv}(\chi, 0)$ is the classical L^2-norm of χ. The computation of successive derivatives is very sensitive numerically. In order to override this numerical stability problem, we can use a B-spline ([deB78] or [S81]) approximation for the curves. Once we have obtained an analytical B-spline expansion for each curve, the successive derivatives are directly computed by differentiating several times their analytic form. More precisely, let $\{B_1, \ldots, B_B\}$ be a B-spline basis. We approximate the discretized curve $x_i = {}^t(\chi_i(t_1), \ldots, \chi_i(t_J))$ as follows :

$$\widehat{\beta}_i = (\widehat{\beta}_{i1}, \ldots, \widehat{\beta}_{iB}) = \arg \inf_{(\alpha_1, \ldots, \alpha_B) \in \mathbb{R}^B} \sum_{j=1}^{p} \left(\chi_i(t_j) - \sum_{b=1}^{B} \alpha_b B_b(t_j) \right)^2 dt.$$

This produces a good approximation of the solution of the minimization problem

$$\arg \inf_{(\alpha_1, \ldots, \alpha_B) \in \mathbb{R}^B} \int \left(\chi_i(t) - \sum_{b=1}^{B} \alpha_b B_b(t) \right)^2 dt,$$

as soon as the grid is sufficiently fine. Hence, we have the following approximation for $x_i = {}^t(\chi_i(t_1), \ldots, \chi_i(t_J))$:

$$\widehat{\chi}_i(.) = \sum_{b=1}^{B} \widehat{\beta}_{ib} B_b(.)$$

As stated above, because the analytic expression of the B_b's is well-known, the successive derivatives can be exactly computed (see again [deB78] or [S81]) and we can differentiate easily the approximated curves :

$$\widehat{\chi}_i^{(q)}(.) = \sum_{b=1}^{B} \widehat{\beta}_{ib} B_b^{(q)}(.),$$

where the analytic expression of $B_b^{(q)}(.)$ is known too. Now, given two discretized curves x_i and $x_{i'}$, we have to compute the following quantities :

$$d_q^{deriv}(x_i, x_{i'}) = \sqrt{\int \underbrace{\left(\widehat{\chi}_i^{(q)}(t) - \widehat{\chi}_{i'}^{(q)}(t) \right)^2}_{f(t)} dt}.$$

It remains to compute the integral which can be done by using the Gauss method (see [L56]). This method can be applied if we are able to evaluate the integrand f at any points, and that is the case here. In fact, the Gauss method proposes to do the following approximation:

$$\int_a^b f(t)dt \sim \frac{b-a}{2} \sum_{k=1}^{K} w_k f\left(\frac{b-a}{2} + \frac{b-a}{2}\delta_k\right)$$

where the weights w_k and the reals δ_k ("gauss points") are tabulated (see [H75]). The accuracy of this numerical method comes from the fact that this method is exact for any polynom of degree $\leq 2K-1$.

To conclude, let us make some remarks about the design points. Because the curves are replaced by their B-spline expansion, the measure $d_q^{deriv}(\chi_i, \chi_{i'})$ does not depend on the grids of discretization (instants of measure), and finally we can use this semi-metric even in the context of unbalanced datasets (i.e., when the curves are not necessarily observed at the same points), as long of course as the grid is fine enough. This looks an appealing point for this kind of semi-metric but in counterpart their use supposes that the f.r.v. are not too rough. Therefore, in practice, this class of semi-metrics will be well adapted and used in the presence of smooth curves, such as the spectrometric ones described before.

3.5 R and S+ Implementations

We have written R and S+ routines for computing the three previously discussed classes of semi-metrics respectively called `semimetric.pca`, `semimetric.mplsr` and `semimetric.deriv`. For more details concerning the sources (code) and guidelines for practical use please see the companion website.[1] Of course, this website can be updated in order to include other semi-metrics not discussed in this book (in particular, the website proposes the routine `semimetric.fourier` based on Fourier expansion).

3.6 What About Unbalanced Functional Data?

There are many situations where we can get unbalanced design points (the location and/or the number of the measurement points can change from one unit to another one). To illustrate, we can consider the following observed sample of curves

$$\chi_i(t_{i,1}), \ \chi_i(t_{i,2}), \ldots, \ \chi_i(t_{i,J_i}),$$

---

[1] *http://www.lsp.ups-tlse.fr/staph/npfda*

for $i = 1, \ldots, n$. This notation indicates clearly that we have at hand n different design points $\mathcal{T}_i = \{t_{i,J_i}\}_{j=1,\ldots,J_i}$ (i.e., one design per unit). Such a situation can be encountered when the starting point of the measurements can differ from a unit to another one (see the *Pinch force* data in Section 1.4.2 of [RS97], the *fda handwriting* data in [KLMR00], or [AOV99b]). Missing data can also lead to irregularly spaced design. An extreme case of unbalanced functional data is the so-called functional sparse data context: the individuals are observed at a sparse set of time points (see the *Spinal bone mineral density* or *Globular filtration rate* examples in [JH01] and [JS03]). According to the unbalanced data setting, specific preliminary processing before any statistical studies have been developed (see [S95] for estimating a shift parameter acting on the design points, [KG92], [GK95] and [KLMR00] for time-warping method in a curve registration context, [JHS00] and [YMW05] for sparse functional data and [B03] for automatic landmark registration).

Here, we consider that we have at hand standard unbalanced sampled functional data in the sense that curve registration and sparse data are out of the scope of this book. Of course, unbalanced functional data can be obtained throughout a preliminary curve registration procedure if it is necessary (see again [RS97] for methodological aspects and [RS02] for case studies). So, we consider an unbalanced functional dataset

$$\{\chi_i(t_{i,j}); \ j = 1, \ldots, J_i\}_{i=1,\ldots,n}$$

such that:

- the n design points $\{\mathcal{T}_i\}_{i-1,\ldots,n}$ (one design by individual) are sufficiently fine (i.e., $max_{i,j}|t_{i,j} - t_{i,j+1}|$ is small enough), which avoids the setting of sparse data,
- locations and/or number of the design points differ from an individual to another one, recalling that this situation does not correspond to a curve registration problem. For instance, if the log-periodograms of the phoneme data described in Section 2.2 were not all sampled at the same frequencies, we would obtain such unbalanced functional data.

It is important to remark that it is always possible to transform an unbalanced functional dataset

$$\{\chi_i(t_{i,j}); \ j = 1, \ldots, J_i\}_{i=1,\ldots,n}$$

into an equally spaced balanced one in a very simple way:

$$\{\widetilde{\chi}_i(t_j); \ j = 1, \ldots, J\}_{i=1,\ldots,n} \ \text{with} \ t_j = t_1 + (j-1)\frac{t_J - t_1}{J - 1}.$$

In the companion website[2], we include the $R/S+$ routine `unbal2equabal` which achieves such a transformation by using a linear interpolation via the $R/S+$ routine `approx`.

---

[2] *http://www.lsp.ups-tlse.fr/staph/npfda*

3.7 Semi-Metric Space: a Well-Adapted Framework

To conclude this chapter one could say that the semi-metric modelization, despite its rather terrifying theoretical look, could be a reasonable way to model functional data. We will see throughout the remainder of the book that such a modelling will allow us both to provide interesting theoretical advances and to give quite appealing results on several functional real datasets. At this stage, one could say that the dataset itself should be a prominent element for choosing which will be the semi-metric to be used. Each of the three families discussed above is adapted to some special kind of data: pca-type semi-metrics are expected to give interesting results for rough datasets, pls-type ones are recommended when one has at hand a multivariate response while derivatives-type ones are adapted to smooth datasets. These ideas will be confirmed throughout the rest of this book. In order to simplify our purpose this monograph deals mainly with balanced and uniformly discretized curves data, but any unbalanced curves data can be treated after some preliminary processing (as indicated in Section 3.6).

4

Local Weighting of Functional Variables

In the finite dimensional case, the local weighting techniques are very popular in the community of nonparametricians because they are very well adapted to nonparametric models. The aim of this chapter is to explain how the concept of local smoothing can be extended to the functional data case. Clearly, local approaches need to have at hand some topological ways for measuring proximity between functional data, and therefore this chapter will be directly linked with the semi-metric modelling discussed in Chapter 3.

In the finite dimensional case, one of the most common approaches among these local weighting methods is certainly the kernel one. It is impossible to give an exhaustive bibliography about nonparametric methods for finite dimensional variables, but the state of art in this field is well summarized in [S00] and [AP03] while a large number of references can be found in [SV00] concerning the kernel methods especially. We will see in this chapter how kernel smoothing ideas can be adapted to infinite dimensional variables.

The chapter is organized as follows. In Section 4.1 we give a basic discussion on kernel method, explaining how (and why) what is classically done for finite dimensional variables can be adapted to functional setting. The second aim (Section 4.2) consists in seeing how the local weighting is in relation to the notion of small ball probabilities. As we will see, small ball probabilities can be viewed as a tool for describing some local behaviours of functional data and the kernel approach allows us to take into account these kinds of local properties. Section 4.3 proposes some general theoretical aspects concerning kernel weighting. Finally, note that Sections 4.1 and 4.2 is of interest to a large public whereas Section 4.3 is meant for statisticians interested in theoretical aspects.

4.1 Why Use Kernel Methods for Functional Data?

Kernel methods are well-known and intensively used by the community of nonparametricians because they are a useful way to do local weighting. We start by recalling shortly what is kernel local weighting in the real and multivariate cases before extending it to the functional context.

4.1.1 Real Case

As it is well known, kernel local weighting is based on a kernel function (classically denoted by K) and on a smoothing parameter, which is called bandwidth and usually denoted by h. If x is a fixed real number, the kernel local weighting transforms n r.r.v. $X_1, X_2, \ldots, X_n$ into $\Delta_1, \Delta_2, \ldots, \Delta_n$ such that:

$$\Delta_i = \Delta_i(x, h, K) = \frac{1}{h} K \left(\frac{x - X_i}{h} \right).$$

The main idea of the local weighting around x is to attribute at each r.r.v. X_i a weight taking into account the distance between x and X_i; the more X_i is distant from x, the smaller is the weighting.

Before going on, let us recall what is a kernel function exactly in this simplest situation. In fact, there exists a large variety of kernels. Any density function can be considered as a kernel, but even unnecessary positive functions can be used ([GM84]). A large literature exists on this field (see [MN89] and [B93] for interesting advances and [HVZ] for a presentation of the state of art). To simplify our purpose, we consider at this stage only positive and symmetrical kernels which are the most classical ones. Figure 4.1 displays various kernel functions which are analytically defined as follows:

(a) *Box* kernel: $K(u) = \frac{1}{2} 1_{[-1,+1]}(u)$,

(b) *Triangle* kernel: $K(u) = (u + 1)1_{[-1,0]}(u) + (1 - u)1_{[0,+1]}(u)$,

(c) *Quadratic* kernel: $K(u) = \frac{3}{4}(1 - u^2)1_{[-1,+1]}(u)$,

(d) *Gaussian* kernel: $K(u) = \frac{1}{\sqrt{2\pi}} \exp\{-\frac{u^2}{2}\}$.

To precise the notion of kernel local weighting, let us consider the Box kernel and rewrite the Δ_i's as follows:

$$\Delta_i = \frac{1}{h} 1_{[x-h,x+h]}(X_i).$$

In this situation, the local feature of the weighting appears obvious since the r.r.v. outside the range $[x-h, x+h]$ are ignored. In addition, the normalization $1/h$ is proportional to the size of the set $[x - h, x + h]$ on which the X_i's are taken into account. These points are not only true for the Box kernel, but are shared by any compactly supported kernels.

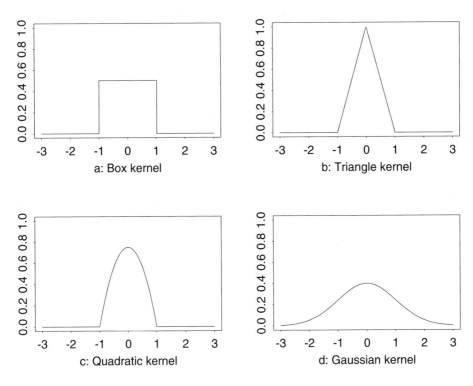

Fig. 4.1. Usual Symmetrical Kernels

4.1.2 Multivariate Case

In multivariate situations one is observing n random vectors $\boldsymbol{X}_1, \boldsymbol{X}_2, \ldots, \boldsymbol{X}_n$ valued in $\mathbb{R}^p$. The previous kernel local weighting can be extended easily to this situation. To that end, it suffices to consider a multivariate kernel K^* which will be a function from $\mathbb{R}^p$ into $\mathbb{R}$. The first (natural) way to do that is to define K^* as a product of p real kernel functions $K_1, \ldots, K_p$:

$$\forall \boldsymbol{u} = {}^t(u_1, \ldots, u_p) \in \mathbb{R}^p, K^*(\boldsymbol{u}) = K_1(u_1) \times K_2(u_2) \times \cdots \times K_p(u_p).$$

As pointed out in [HM00], a second way consists in combining a real kernel function H with a norm (denoted by $\|.\|$) in $\mathbb{R}^p$ as follows:

$$\forall \boldsymbol{u} \in \mathbb{R}^p, K^*(\boldsymbol{u}) = K(\|\boldsymbol{u}\|).$$

Note that if $K_1 = K_2 = \cdots = K_p = 1_{[-1,1]}$ and if $\|.\|$ is the supremum norm, both approaches coincide by taking $K = 1_{[0,1]}$. Moreover, because $\|\boldsymbol{u}\|$ is always a positive quantity, the real kernel K should have a positive support (i.e., $\{v \in \mathbb{R}$ such that $K(v) > 0\} \subset \mathbb{R}^+$). This leads to use asymmetrical functions for the kernel K. The few examples reported in Figure 4.2 are the asymmetrical versions of the ones displayed in Figure 4.1.

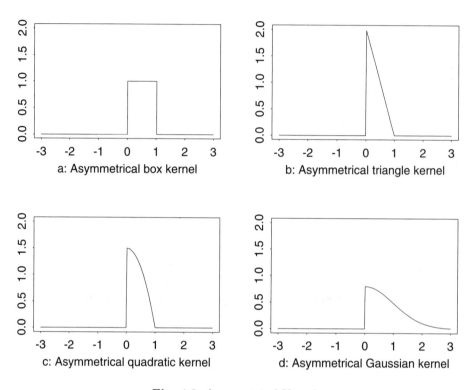

Fig. 4.2. Asymmetrical Kernels

Now, let us discuss how this can be interpreted in terms of local weighting. Indeed, what happens is very similar to the real case. Let x be a fixed vector of $\mathbb{R}^p$. The multivariate kernel local weighting consists in transforming the n random vectors $X_1, X_2, \ldots, X_n$ into the n variables $\Delta_1, \Delta_2, \ldots, \Delta_n$:

$$\Delta_i = \frac{1}{h^p} K^* \left(\frac{x - X_i}{h} \right).$$

If we consider compactly supported kernels, it appears clearly that the Δ_i are locally weighted transformations of the variables X_i, since $\Delta_i = 0$ as long as the corresponding X_i is out of some neighborhood of x. Moreover, the normalization $1/h^p$ is proportional to the volume of the set on which the X_i's are taken into account.

4.1.3 Functional Case

The background presented above is sufficient to introduce the kernel local weighting in the functional case. Let $\mathcal{X}_1, \mathcal{X}_2, \ldots, \mathcal{X}_n$ be n f.r.v. valued in E and let χ be a fixed element of E. A naive functional extension of multivariate kernel local weighting ideas would be to transform the n f.r.v. $\mathcal{X}_1, \mathcal{X}_2, \ldots, \mathcal{X}_n$ into the n quantities

$$\frac{1}{V(h)} K \left(\frac{d(\chi, \mathcal{X}_i)}{h} \right),$$

where d is a semi-metric on E, K is a real (asymmetrical) kernel. In this expression $V(h)$ would be the volume of

$$B(\chi, h) \; = \; \{\chi' \in E, \; d(\chi, \chi') \leq h\}$$

which is the ball, with respect to the topology induced by d, centered at χ and of radius h. However, this naive approach requests to define $V(h)$. In other words, this needs to have at hand a measure on E. This is the main difference with real and multivariate cases for which the Lebesgue measure is implicitly used whereas in the functional space E we do not have such a universally accepted reference measure (see [D02] for deeper discussion). Therefore, in order to free oneself of a choice of a particular measure, we build the normalization by using directly the probability distribution of the f.r.v. The *functional* kernel local weighted variables are defined by:

$$\Delta_i \; = \; \frac{K \left(\frac{d(\chi, \mathcal{X}_i)}{h} \right)}{\mathbb{E} \left(K \left(\frac{d(\chi, \mathcal{X}_i)}{h} \right) \right)}. \tag{4.1}$$

If we go back quickly to the multivariate case we have, for some constant C depending on K and on the norm $||.||$ used in $\mathbb{R}^p$,

$$\mathbb{E} K \left(||\boldsymbol{x} - \boldsymbol{X}_i|| / h \right) \sim C f(\boldsymbol{x}) h^p \tag{4.2}$$

as long as $\boldsymbol{X}_i$ has a density f with respect to Lebesgue measure which is continuous and such that $f(\boldsymbol{x}) > 0$ (this kind of result is known in the literature as the Bochner's type theorem and [C76] gives a large scope on such results). So, it is clear now that (4.1) is an extension of the multivariate kernel local weighting in the functional framework.

Note that the kernel functions K to be used here are necessarily the asymmetrical ones described in Section 4.1.2 above. For the sake of simplicity, in the remainder of this book, we will consider only two kinds of kernels for weighting functional variables.

Definition 4.1. *i) A function K from $\mathbb{R}$ into $\mathbb{R}^+$ such that $\int K = 1$ is called a kernel of type I if there exist two real constants $0 < C_1 < C_2 < \infty$ such that:*

$$C_1 1_{[0,1]} \leq K \leq C_2 1_{[0,1]}.$$

ii) A function K from $\mathbb{R}$ into $\mathbb{R}^+$ such that $\int K = 1$ is called a kernel of type II if its support is $[0,1]$ and if its derivative K' exists on $[0,1]$ and satisfies for two real constants $-\infty < C_2 < C_1 < 0$:

$$C_2 \leq K' \leq C_1.$$

The first kernel family contains the usual discontinuous kernels such as the asymmetrical box one while the second family contains the standard asymmetrical continuous ones (as the triangle, quadratic, ...). Finally, to be in harmony with this definition and simplify our purpose, for local weighting of real random variables we just consider the following kernel-type.

Definition 4.2. *A function K from $\mathbb{R}$ into $\mathbb{R}^+$ such that $\int K = 1$ with compact support $[-1,1]$ and such that $\forall u \in (0,1)$, $K(u) > 0$ is called a kernel of type 0.*

4.2 Local Weighting and Small Ball Probabilities

We can now build the bridge between local weighting and the notion of small ball probabilities. To fix the ideas, consider the simplest kernel among those of type I namely the asymmetrical box kernel. Let $\mathcal{X}$ be a f.r.v. valued in E and χ be again a fixed element of E. We can write:

$$\mathbb{E}\left(1_{[0,1]}\left(\frac{d(\chi, \mathcal{X})}{h}\right)\right) = \mathbb{E}\left(1_{B(\chi,h)}(\mathcal{X})\right) = P\left(\mathcal{X} \in B(\chi,h)\right). \qquad (4.3)$$

Keeping in mind the functional kernel local weighted variables (4.1), the probability of the ball $B(\chi, h)$ appears clearly in the normalization. At this stage it is worth telling why we are saying *small* ball probabilities. In fact, as we will see later on, the smoothing parameter h (also called the *bandwidth*) decreases with the size of the sample of the functional variables (more precisely, h tends to zero when n tends to ∞). Thus, when we take n very large, h is close to zero and then $B(\chi, h)$ is considered as a small ball and $P\left(\mathcal{X} \in B(\chi, h)\right)$ as a small ball probability.

From now on, for all χ in E and for all positive real h, we will use the notation:

$$\varphi_\chi(h) = P\left(\mathcal{X} \in B(\chi, h)\right). \tag{4.4}$$

This notion of small ball probabilities will play a major role both from theoretical and practical points of view. Because the notion of ball is strongly linked with the semi-metric d, the choice of this semi-metric will become an important stage. This will be seen from a theoretical point of view throughout the book, since the rates of convergence of our nonparametric functional estimates will be systematically linked with d through the behaviour, around 0, of the small ball probability function φ_χ. In addition, let us recall that the discussion in Chapter 3 has emphasized the fact that the choice of the semi-metric is also expected to be a crucial point as long as we focus on the applied aspects.

4.3 A Few Basic Theoretical Advances

Because the functional kernel local weighting ideas will be at the heart of all the functional nonparametric statistical methods to be studied later in this book, we decided to put together here some short common results. We will state two results, according to the fact that the kernel is of type I or II, that can be seen as functional versions of what is known as the Bochner theorem in the usual nonparametric finite dimensional literature (see (4.2) and discussion). Naturally, as the Bochner theorem does in finite dimension, both of the following lemmas will be used very often throughout the rest of this book. Before going on, let $\mathcal{X}$ be a f.r.v. taking its values in the semi-metric space (E, d), let χ be a fixed element of E, let h be a real positive number and let K be a kernel function.

Lemma 4.3. *If K is a kernel of type I, then there exist nonnegative finite real constants C and C' such that:*

$$C\,\varphi_\chi(h) \le \mathbb{E}\,K\left(\frac{d(\chi, \mathcal{X})}{h}\right) \le C'\,\varphi_\chi(h). \tag{4.5}$$

Proof. Because K is a kernel of type I, we have by Definition 4.1i,

$$C_1 1_{[0,1]} \le K \le C_2 1_{[0,1]},$$

which implies directly that

$$C_1 1_{B(\chi,h)}(\mathcal{X}) \le K\left(\frac{d(\chi, \mathcal{X})}{h}\right) \le C_2 1_{B(\chi,h)}(\mathcal{X}).$$

It suffices to apply (4.3) in order to obtain the claimed result with $C = C_1$ and $C' = C_2$. $\square$

Lemma 4.4. *If K is a kernel of type II and if $\varphi_\chi(.)$ satisfies*

$$\exists C_3 > 0, \exists \epsilon_0, \forall \epsilon < \epsilon_0, \int_0^\epsilon \varphi_\chi(u)du > C_3 \epsilon \varphi_\chi(\epsilon), \qquad (4.6)$$

then there exist nonnegative finite real constants C and C' such that, for h small enough:

$$C\,\varphi_\chi(h) \leq \mathbb{E}\,K\left(\frac{d(\chi, \mathcal{X})}{h}\right) \leq C'\,\varphi_\chi(h). \qquad (4.7)$$

Proof. We start by writing

$$\mathbb{E}\,K\left(\frac{d(\chi, \mathcal{X})}{h}\right) = \int_0^1 K(t)\,dP^{\frac{d(\chi, \mathcal{X})}{h}}(t),$$

and because K' exists, we have $K(t) = K(0) + \int_0^t K'(u)\,du$, which implies that

$$\mathbb{E}\,K\left(\frac{d(\chi, \mathcal{X})}{h}\right) = \int_0^1 K(0)\,dP^{\frac{d(\chi, \mathcal{X})}{h}}(t) + \int_0^1 \left(\int_0^t K'(u)\,du\right) dP^{\frac{d(\chi, \mathcal{X})}{h}}(t),$$

$$= K(0)\,\varphi_\chi(h) + \int_0^1 \left(\int_0^1 K'(u)\,1_{[u,1]}(t)\,du\right) dP^{\frac{d(\chi, \mathcal{X})}{h}}(t),$$

$$= K(0)\,\varphi_\chi(h) + \int_0^1 K'(u)\,P\left(u \leq \frac{d(\chi, \mathcal{X})}{h} \leq 1\right) du,$$

the last equation being obtained by applying Fubini's Theorem. Using the fact that $K(1) = 0$ allows us to write

$$\mathbb{E}\,K\left(\frac{d(\chi, \mathcal{X})}{h}\right) = -\int_0^1 K'(u)\,\varphi_\chi(hu)\,du. \qquad (4.8)$$

It suffices to use (4.6) to show that, for $h < \epsilon_0$ and with $C = -C_3\,C_1$,

$$\mathbb{E}\,K\left(\frac{d(\chi, \mathcal{X})}{h}\right) \geq C\,\varphi_\chi(h).$$

Concerning the upper bound, it suffices to remark that K is bounded with support $[0, 1]$ and the same arguments as for Lemma 4.3 can be used by putting $C' = \sup_{t \in [0,1]} K(t)$. $\square$

Hypothesis (4.6) allows us to control more precisely the behaviour of the small ball probability function $\varphi_\chi(.)$ around zero. We will see in the final steps of this book (see Remark 13.3 and Lemma 13.6), that it is always possible to choose a semi-metric $d(.,.)$ for which this condition is fulfilled.

Part II

Nonparametric Prediction from Functional Data

A well-known statistical problem consists in studying the link between two variables in order to predict one of them (the response variable) given a new value for the other one (the explanatory variable). This problem has been widely studied for real or multivariate variables, but it also obviously occurs with functional variables. This part of the book is devoted to this problem when the response variable is real and the explanatory variable is functional. This problem is called *prediction of scalar response from functional variable*. We describe some recent statistical ideas, based on an independent statistical sample (dependent extensions are discussed in Part IV).

There are several ways to approach the prediction setting, and one of the most popular is certainly the regression method which is based on conditional expectation. For robustness purposes (according to the behavior of the conditional distribution) two alternative techniques are *conditional median* and *conditional mode*. These three predictors have been widely studied in multivariate situations, and our aim is to adapt them to the situation where the explanatory variables are possibly of infinite dimension. In the functional framework, each of these three functional prediction approaches leads to the estimation of some (non-linear) operator: the conditional expectation (for functional regression), the conditional cumulative distribution function (for functional median) and the conditional density (for functional mode). In order to keep the "free-modelling" feature and to avoid too restrictive modelling assumptions, we attack these problems by means of nonparametric models for these (non-linear) operators. As discussed in Chapter 4, kernel local weighting is well adapted for this purpose, and all our nonparametric estimates will be based on these kernel ideas. The main goal of this part is to propose a systematic study of each previous functional nonparametric prediction method by developing both practical and theoretical aspects.

This part is organized as follows: Chapter 5 describes the three prediction methods and the three associated nonlinear functional operators, with special attention to the statement of the associated statistical models and estimators. Chapter 6 gives mathematical support by giving consistency results for each nonparametric functional estimate. The rates of convergence will be linked both with the nonparametric model and with the semi-metric and this last point is directly connected with the small ball probabilities of the functional variable. In this sense, Chapter 6 gives theoretical support to some empirical ideas pointed out before in Chapter 3. Then, Chapter 7 highlightes the practical issues, including as well application to the spectrometric dataset discussed in Chapter 2 as computational features. For each prediction method some R and S+ routines are given.

This field of research is quite new, and both theoretical and practical advances give appealing results. So, we are definitively supporting the idea that many interesting further developments on functional nonparametric prediction should (and will) born in the future, and we leave an important place in our presentation for to the statement of open problems.

5

Functional Nonparametric Prediction Methodologies

This chapter describes several approaches concerning the nonparametric prediction of some scalar response. The functional setting appears through the explanatory functional variable. We focus on three complementary prediction methods, namely the conditional expectation, the conditional median and the conditional mode. Conditional expectation refers to the well-known regression method whereas both the conditional median and conditional mode are strongly linked with the estimation of the conditional distribution. After introducing some bibliographical aspects (Section 5.1), we present in Section 5.2 the three functional nonparametric prediction methods. Section 5.3 presents the nonparametric models associated with these prediction problems, while Section 5.4 focuses on the construction of the estimators.

5.1 Introduction

There are many situations in which one may wish to study the link between two variables, with the main goal to be able to predict new values of one of them given the other one. This prediction problem has been widely studied in the literature when both variables are of finite dimensions. Of course the same problem can occur when some of the variables are functional. Our wish is to investigate this problem when the explanatory variable is functional and the response one is still real. Both to fix the ideas and to emphasize the great interest and usefulness of this problem in many fields of applied sciences, let us quickly come back to the chemometric data presented in Chapter 2.

As discussed in Section 2.1, recall that the statistical sample (of size $n = 215$) is composed of spectrometric curves $\chi_1, \ldots, \chi_n$ (these are the functional data) corresponding to the spectra observed for 215 pieces of finely chopped meat. In addition, by an analytic chemical process we have measured the fat content of each piece $y_1, \ldots, y_n$ (these are the scalar responses). Thus, we collect the observations of a scalar response (the fat content) and an explanatory functional variable (spectra). One question is: given an observed spectrum of a piece of meat, can we predict its corresponding fat content? This is typically a functional prediction problem. To answer the question, we have to estimate the link between the fat content and the spectra. Unfortunately, there is neither a way to display this relationship nor is there structural information about it. Therefore, it becomes natural to introduce nonparametric models in order to make as few assumptions as possible on the shape of the link. The functional aspect of the problem is very important too, and we have to attack it in such a way as to use the whole spectrometric curve. In particular, continuity and other functional features of the spectra have to be taken into account. There is therefore real need for developing methods combining both nonparametric concepts and functional variable modelling.

Of course, there is a consistent literature both around nonparametric prediction and functional data. But, until now, functional variables have been studied essentially in a parametric setting. This has been popularized by [RS97] (mainly for practical points of view) and previous theoretical developments can be found in [B00] in the specific context of dependent functional variables. Recent practical advances can be found for instance in [CGS04] whereas some asymptotic studies are detailed in [CFS03] and [CFF02].

In another direction of statistical research, nonparametric prediction problems have been investigated intensively both in real and multivariate cases. It is impossible to give an exhaustive description of the related bibliography, but to fix the ideas the reader could look at the precursor works by [W64] and [N64], at the intermediary survey by [C85] and at [S00] or [AP03] for a description of the state of the art.

The aim of this book is to marry advantages of free-modelling together with a fully functional methodology in order to answer to functional prediction problem such as the spectrometric one. In this chapter, we present three functional nonparametric statistical approaches for the prediction problem. The reader has to keep in mind Defintion 1.4 and the fact that in the designation *functional nonparametric prediction method*, the word *functional* refers to the concept of functional variable (implicitly "we have to take into account the functional feature of the variable") while the word *nonparametric* means that we use a free-parameter modelling for the nonlinear operators to be estimated. In addition, it is important to note that our methodology is also based on free-distribution modelling since no parametric assumption is necessary for the distribution of the random variables.

5.2 Various Approaches to the Prediction Problem

Let us start by recalling some notation. Let $(\mathcal{X}_i, Y_i)_{i=1,\ldots,n}$ be n independent pairs, identically distributed as $(\mathcal{X}, Y)$ and valued in $E \times \mathbb{R}$, where (E, d) is a semi-metric space (i.e. $\mathcal{X}$ is a f.r.v. and d a semi-metric). Let χ (resp. y) be a fixed element of E (resp. $\mathbb{R}$), let $\mathcal{N}_\chi \subset E$) be a neighboorhood of χ and S

be a fixed compact subset of $\mathbb{R}$. Given χ, let us denote by $\widehat{y}$ a predicted value for the scalar response.

We propose to predict the scalar response Y from the functional predictor $\mathcal{X}$ by using various methods all based on the conditional distribution of Y given $\mathcal{X}$. This leads naturally to focus on some conditional features such as conditional expectation, median and mode. The regression (nonlinear) operator r of Y on $\mathcal{X}$ is defined by

$$r(\chi) = \mathbb{E}(Y|\mathcal{X} = \chi), \tag{5.1}$$

and the conditional cumulative distribution function (c.d.f.) of Y given $\mathcal{X}$ is defined by:

$$\forall y \in \mathbb{R}, \; F_Y^{\mathcal{X}}(\chi, y) = P(Y \le y|\mathcal{X} = \chi). \tag{5.2}$$

In addition, if the probability distribution of Y given $\mathcal{X}$ is absolutely continuous with respect to the Lebesgue measure, we note $f_Y^{\mathcal{X}}(\chi, y)$ the value of the corresponding density function at (χ, y). Note that under a differentiability assumption on $F_Y^{\mathcal{X}}(\chi, .)$, this functional conditional density can be written as

$$\forall y \in \mathbb{R}, \; f_Y^{\mathcal{X}}(\chi, y) = \frac{\partial}{\partial y} F_Y^{\mathcal{X}}(\chi, y). \tag{5.3}$$

For these two last definitions, we are implicitly assuming that there exists a regular version of this conditional probability. In the remainder of this book, this assumption will be done implicitly as long as we will need to introduce this conditional cdf $F_Y^{\mathcal{X}}(\chi, y)$ or the conditional density $f_Y^{\mathcal{X}}$. It is out of the scope of this book to discuss probabilistic conditions insuring such an existence. Let us just recall that if d is a metric, such an existence is insured under quite general separability assumptions (see for instance [J84], or [T63]) while for more general space this is still a field of actual probabilistic researches (see for instance [F85], [M85] or [LFR04] for recent advances and references). This point occurs also in other fields of statistics, such as for instance inverse problems (see [L89]).

It is clear that each of these nonlinear operators gives information about the link between $\mathcal{X}$ and Y and thus can be useful for predicting y given χ. Indeed, each of them will lead to some specific prediction method. The first way to construct such a prediction is obtained directly from the regression operator by putting:

$$\widehat{y} = \widehat{r}(\chi), \tag{5.4}$$

$\widehat{r}$ being an estimator of r. The second one consists of considering the median $m(\chi)$ of the conditional c.d.f. $F_Y^{\mathcal{X}}$:

$$m(\chi) = \inf \left\{ y \in \mathbb{R}, \; F_Y^{\mathcal{X}}(\chi, y) \ge 1/2 \right\}, \tag{5.5}$$

and to use as predictor:

$$\widehat{y} = \widehat{m}(\chi), \tag{5.6}$$

where $\widehat{m}(\chi)$ is an estimator of this functional conditional median $m(\chi)$. Note that such a conditional median estimate will obviously depend on some previous estimation of the nonlinear operator $F_Y^{\mathcal{X}}$. Finally, the third predictor is based directly on the mode $\theta(\chi)$ of the conditional density of Y given $\mathcal{X}$:

$$\theta(\chi) = \arg\sup_{y \in S} f_Y^{\mathcal{X}}(\chi, y). \tag{5.7}$$

This definition assumes implicitly that $\theta(\chi)$ exists on S. The predictor is defined by:

$$\widehat{y} = \widehat{\theta}(\chi), \tag{5.8}$$

where $\widehat{\theta}(\chi)$ is an estimator of this functional conditional mode $\theta(\chi)$. Once again note that this conditional mode estimate will directly depend on some previous estimation of the nonlinear operator $f_Y^{\mathcal{X}}$.

What about predictive confidence band? Indeed, the three methods presented before are concerning pointwise prediction. Nevertheless, it is worth to note that the second method can also be used for prediction confidence band construction, since it can be used for estimating any quantile of the conditional c.d.f. $F_Y^{\mathcal{X}}$. Precisely, these quantiles are defined for $\alpha \in (0,1)$ by:

$$t_\alpha(\chi) = \inf\left\{ y \in \mathbb{R}, \ F_Y^{\mathcal{X}}(\chi, y) \geq \alpha \right\}. \tag{5.9}$$

Thus, from an estimate $\widehat{t}_\alpha(\chi)$ of $t_\alpha(\chi)$, the following interval

$$\left[\widehat{t}_\alpha(\chi), \ \widehat{t}_{1-\alpha}(\chi) \right] \tag{5.10}$$

is one way to build, for $\alpha \in (0, 1/2)$, a $(1 - 2\alpha)$ predictive confidence band.

5.3 Functional Nonparametric Modelling for Prediction

At this stage, it remains to construct explicitly some estimates of the theoretical predictors which have been introduced (that is, regression, conditional median, conditional mode and conditional quantiles). However, this cannot be done before having precisely the kind of statistical model we wish to introduce. That is the aim of this section.

Each of the three predictors is based on the estimation of some (nonlinear) operator: the regression operator r, the conditional c.d.f. $F_Y^{\mathcal{X}}$ or its density function $f_Y^{\mathcal{X}}$. Therefore, the first stage of the statistical modelling consists of introducing some sets of constraints acting either on r, $F_Y^{\mathcal{X}}$ or $f_Y^{\mathcal{X}}$. Keeping in mind that we wish to find free-parameter models, this leads to the introduction of nonparametric models (according to Definition 1.4). In the following, we only consider two kinds of nonparametric models: models based on a continuity-type condition and models based on Lipschitz-type condition. To fix the ideas, we decided to put all the nonparametric models together in this section.

Prediction via conditional expectation. This method referes to the regression operator r (which is a nonlinear operator from E to $\mathbb{R}$) and we will consider the following models:

$$r \in C_E^0, \tag{5.11}$$

where

$$C_E^0 = \left\{ f : E \to \mathbb{R}, \lim_{d(\chi,\chi') \to 0} f(\chi') = f(\chi) \right\},$$

or $\exists \beta > 0$ such that

$$r \in Lip_{E,\beta}, \tag{5.12}$$

where

$$Lip_{E,\beta} = \left\{ f : E \to \mathbb{R}, \exists C \in \mathbb{R}_*^+, \forall \chi' \in E, |f(\chi) - f(\chi')| < C\, d(\chi,\chi')^\beta \right\}.$$

Prediction via functional conditional median. This method referes to the operator $F_Y^{\mathcal{X}}$ (which is a nonlinear operator from $E \times \mathbb{R}$ to $\mathbb{R}$). Before writing the model, in order to simplify the presentation and to not mask our main purpose, we decided to assume that the conditional median operator is lying in the following set

$$\mathcal{S}_{cdf}^{\chi} = \{ f : E \times \mathbb{R} \to \mathbb{R}, f(\chi,.) \text{ is a strictly increasing c.d.f.} \}. \tag{5.13}$$

Indeed, $F_Y^{\mathcal{X}} \in \mathcal{S}_{cdf}^{\chi}$ insures the existence and unicity of the conditional median which can be defined as follows:

$$m(\chi) = F_Y^{\chi^{-1}}(1/2) \text{ where } F_Y^{\chi} = \begin{cases} \mathbb{R} \to [0,1] \\ y \mapsto F_Y^{\chi}(y) = F_Y^{\mathcal{X}}(\chi,y). \end{cases} \tag{5.14}$$

Now, consider the following set of constraints:

$$C_{E \times \mathbb{R}}^0 = \left\{ \begin{array}{c} f : E \times \mathbb{R} \to \mathbb{R}, \forall \chi' \in \mathcal{N}_\chi, \\ \\ \lim_{d(\chi,\chi') \to 0} f(\chi',y) = f(\chi,y) \\ \\ \text{and } \forall y' \in \mathbb{R}, \lim_{|y'-y| \to 0} f(\chi,y') = f(\chi,y) \end{array} \right\}, \tag{5.15}$$

or

$$Lip_{E \times \mathbb{R}, \beta} = \left\{ \begin{array}{c} f : E \times \mathbb{R} \to \mathbb{R}, \\ \\ \forall (\chi_1, \chi_2) \in \mathcal{N}_\chi^2, \forall (y_1, y_2) \in S^2, \\ \\ |f(\chi_1,y_1) - f(\chi_2,y_2)| \leq C \left(d(\chi_1,\chi_2)^\beta + |y_1 - y_2|^\beta \right) \end{array} \right\}. \tag{5.16}$$

Two functional nonparametric models for the conditional c.d.f. operator can be defined as follows:

$$F_Y^{\boldsymbol{\chi}} \in C_{E \times \mathbb{R}}^0 \cap \mathcal{S}_{cdf}^{\chi}, \tag{5.17}$$

or, it exists $\beta > 0$ such that

$$F_Y^{\boldsymbol{\chi}} \in Lip_{E \times \mathbb{R}, \beta} \cap \mathcal{S}_{cdf}^{\chi}. \tag{5.18}$$

Prediction via functional conditional mode. This method refers to the operator $f_Y^{\boldsymbol{\chi}}$ (which is a nonlinear operator from $E \times \mathbb{R}$ to $\mathbb{R}$). Recall that S is a fixed compact $S \in \mathbb{R}$. We start by introducing the following set of constraints:

$$\mathcal{S}_{dens}^{\chi} = \left\{ \begin{array}{c} f : E \times \mathbb{R} \to \mathbb{R}, \\ \exists \xi > 0, \exists! y_0 \in S, \ f(\chi, .) \text{ is strictly increasing on } (y_0 - \xi, y_0) \\ \text{and strictly decreasing on } (y_0, y_0 + \xi). \end{array} \right\} \tag{5.19}$$

It is clear that if $f_Y^{\boldsymbol{\chi}}$ lies to $\mathcal{S}_{dens}^{\chi}$, the problem of maximizing $f_Y^{\boldsymbol{\chi}}(\chi, y)$ over S has a unique solution which is exactly y_0. Therefore, under this restriction, the conditional mode $\theta(\chi)$ can be alternatively defined to be such that

$$\theta(\chi) = \arg\sup_{y \in S} f_Y^{\boldsymbol{\chi}}(\chi, y). \tag{5.20}$$

The set of constraints $\mathcal{S}_{dens}^{\chi}$ is a way for insuring the unicity of the conditional mode $\theta(\chi)$. Now, we define two functional nonparametric models for prediction via conditional mode as follows:

$$f_Y^{\boldsymbol{\chi}} \in C_{E \times \mathbb{R}}^0 \cap \mathcal{S}_{dens}^{\chi}, \tag{5.21}$$

or, it exists $\beta > 0$ such that

$$f_Y^{\boldsymbol{\chi}} \in Lip_{E \times \mathbb{R}, \beta} \cap \mathcal{S}_{dens}^{\chi}. \tag{5.22}$$

Functional nonparametric models (5.11), (5.17) and (5.21) will be called *continuity-type* functional nonparametric models because the continuity property is their main common functional feature. Functional nonparametric models (5.12), (5.18) and (5.22) will be called *Lipschitz-type* functional nonparametric models for the same reason. In fact, we will see that the continuity-type models allow us to obtain convergence results for the nonparametric estimates, whereas the Lipschitz-type ones allows the precise rate of convergence to be found. These theoretical considerations are not surprising because they obey the following general statistical principle: the more the model is restrictive the more the asymptotic behavior can be described precisely. In other words, there is a trade-off between the size of the set of constraints (which produces the model) and the accuracy of the rate of convergence that we can expect.

Another important point about such functional nonparametric models is that two main difficulties appear: the infinite dimensional space of constraints (i.e., the nonparametric model) and the infinite dimensional space of E (i.e., the functional feature of the explanatory variable). Therefore, a great theoretical challenge consists in providing asymptotic properties in such a double infinite dimensional context, and this will be done in Chapter 6. Before that, we need to discuss how to construct functional nonparametric estimates well adapted to such kinds of statistical models.

5.4 Kernel Estimators

Once the nonparametric modelling has been introduced, we have to find ways to estimate the various mathematical objects exhibited in the previous models, namely the (nonlinear) operators r , $F_Y^{\mathcal{X}}$ and $f_Y^{\mathcal{X}}$. According to the discussion in Chapter 4, kernel estimators are good candidates for achieving a local weighting approach in the functional setting. As we will see, they combine both of the following advantages: simple expression and ease of implementation.

Estimating the regression. We propose for the nonlinear operator r the following functional kernel regression estimator:

$$\widehat{r}(\chi) \; = \; \frac{\sum_{i=1}^{n} Y_i \, K\left(h^{-1} d(\chi, \mathcal{X}_i)\right)}{\sum_{i=1}^{n} K\left(h^{-1} d(\chi, \mathcal{X}_i)\right)}, \qquad (5.23)$$

where K is an asymmetrical kernel and h (depending on n) is a strictly positive real. It is a functional extension of the familiar Nadaraya-Watson estimate ([N64] and [W64]) which was previously introduced for finite dimensional nonparametric regression (see [H90] for extensive discussion). The main change comes from the semi-metric d which measures the proximity between functional objects. To see how such an estimator works, let us consider the following quantities:

$$w_{i,h}(\chi) \; = \; \frac{K\left(h^{-1} d(\chi, \mathcal{X}_i)\right)}{\sum_{i=1}^{n} K\left(h^{-1} d(\chi, \mathcal{X}_i)\right)}.$$

Thus, it is easy to rewrite the kernel estimator (5.23) as follows:

$$\widehat{r}(\chi) \; = \; \sum_{i=1}^{n} w_{i,h}(\chi) \, Y_i. \qquad (5.24)$$

which is really a weighted average because:

$$\sum_{i=1}^{n} w_{i,h}(\chi) = 1.$$

The behavior of the $w_{i,h}(\chi)$'s can be deduced from the shape of the asymmetrical kernel function K (see for instance Figure 4.2 in Chapter 4).

To fix the ideas, let us consider positive kernels supported on $[0,1]$. In this case, it is clear that the smaller $d(\chi, \boldsymbol{\mathcal{X}}_i)$, the larger $K\left(h^{-1} d(\chi, \boldsymbol{\mathcal{X}}_i)\right)$. In other words, the closer $\boldsymbol{\mathcal{X}}_i$ is to χ, the larger is the weight assigned to Y_i. The local aspect of such a method appears through local behavior of the weights around χ, the point of E at which the estimator is valued. Note also that, as soon as $d(\chi, \boldsymbol{\mathcal{X}}_i) > h$, we have $w_{i,h}(\chi) = 0$. That means that the estimator $\widehat{r}(\chi)$ is only taken into account among Y_i's those for which the corresponding $\boldsymbol{\mathcal{X}}_i$'s are distant from χ of at most h. So, the parameter h plays a major role because it controls the number of terms in the weighted average. Indeed, the smaller h is, the smaller the number of Y_i's taken into account in the average. In other words, the smaller h is, the more $\widehat{r}(\chi)$ is sensitive to small variations of the Y_i's. In the opposite case, the larger h is, the larger the number of terms in the sum and the less sensitive $\widehat{r}(\chi)$ is with respect to small variations of the Y_i's. We can say that h has a smoothing effect, and in this sense h is a *smoothing parameter*. Because h allows us to select the number of the terms contained in the expression of the estimator, this smoothing parameter is also called bandwidth, which is the usual designation in the classical (real or multivariate) nonparametric context.

Estimating the conditional c.d.f.. We focus now on the estimator $\widehat{F}_Y^{\boldsymbol{\mathcal{X}}}$ of the conditional c.d.f. $F_Y^{\boldsymbol{\mathcal{X}}}$, but let us first explain how we can extend the idea previously used for the construction of the kernel regression estimator. Clearly, $F_Y^{\boldsymbol{\mathcal{X}}}(\chi, y) = P(Y \le y | \boldsymbol{\mathcal{X}} = \chi)$ can be expressed in terms of conditional expectation:

$$F_Y^{\boldsymbol{\mathcal{X}}}(\chi, y) = \mathbb{E}\left(1_{(-\infty, y]}(Y) | \boldsymbol{\mathcal{X}} = \chi\right),$$

and by analogy with the functional regression context, a naive kernel conditional c.d.f. estimator could be defined as follows:

$$\widetilde{F}_Y^{\boldsymbol{\mathcal{X}}}(\chi, y) = \frac{\sum_{i=1}^n K\left(h^{-1} d(\chi, \boldsymbol{\mathcal{X}}_i)\right) 1_{(-\infty, y]}(Y_i)}{\sum_{i=1}^n K\left(h^{-1} d(\chi, \boldsymbol{\mathcal{X}}_i)\right)}.$$

By following the ideas previously developed by [Ro69] and [S89] in the finite dimensional case, it is easy to construct a smooth version of this naive estimator. To do so, it suffices to change the basic indicator function into a smooth c.d.f. Let K_0 be an usual symmetrical kernel (see examples in Section 4.1), let H be defined as:

$$\forall u \in \mathbb{R}, \qquad H(u) = \int_{-\infty}^u K_0(v)\, dv, \tag{5.25}$$

and define the kernel conditional c.d.f. estimator as follows:

$$\widehat{F}_Y^{\boldsymbol{\mathcal{X}}}(\chi, y) = \frac{\sum_{i=1}^n K\left(h^{-1} d(\chi, \boldsymbol{\mathcal{X}}_i)\right) H\left(g^{-1}(y - Y_i)\right)}{\sum_{i=1}^n K\left(h^{-1} d(\chi, \boldsymbol{\mathcal{X}}_i)\right)}, \tag{5.26}$$

where g is a strictly positive real number (depending on n). Figure 5.1 gives various examples of integrated kernels built from standard symmetrical ones.

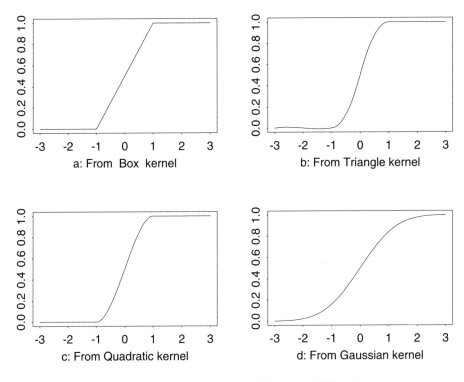

Fig. 5.1. Various Examples of Integrated Kernels

To fix the ideas, let us consider K_0 as a kernel of type 0 (see Definition 4.2). In this case, H is a c.d.f. and the quantity $H\left(g^{-1}\left(y-Y_i\right)\right)$ acts as a local weighting: when Y_i is less than y the quantity $H\left(g^{-1}\left(y-Y_i\right)\right)$ is large and the more Y_i is above y, the smaller the quantity $H\left(g^{-1}\left(y-Y_i\right)\right)$. Moreover, we can write:

$$H\left(g^{-1}\left(y-Y_i\right)\right) = \begin{cases} 0 \Leftrightarrow y \leq Y_i - g, \\ 1 \Leftrightarrow y \geq Y_i + g. \end{cases}$$

It is clear that the parameter g acts as the bandwidth h. The smoothness of the function $\widehat{F}_Y^{\mathcal{X}}(\chi, .)$ is controled both by the smoothing parameter g and by the regularity of the c.d.f. H. The idea to build such a smooth c.d.f. estimate was introduced by [A81] and [R81] (see Section 6.4.2 for a wider bibliographical discussion). The roles of the other parameters involved in this functional kernel c.d.f. estimate (i.e., the roles of K and h) are the same as in the regression setting.

From this conditional c.d.f. estimate (5.26), one can attack the prediction problem by defining a kernel estimator of the functional conditional median

$m(\chi)$ as follows:

$$\widehat{m}(\chi) = \inf \left\{ y \in \mathbb{R}, \ \widehat{F}_Y^{\boldsymbol{\chi}}(\chi, y) \geq \frac{1}{2} \right\}. \tag{5.27}$$

More generally, we can also define from (5.26) a kernel estimator of the functional conditional quantiles $t_\alpha(\chi)$, for any α in $[0, 1/2]$, as follows:

$$\widehat{t}_\alpha(\chi) = \inf \left\{ y \in \mathbb{R}, \ \widehat{F}_Y^{\boldsymbol{\chi}}(\chi, y) \geq \alpha \right\}. \tag{5.28}$$

Estimating the conditional density. It is known that, under some differentiability assumption, the conditional density function can be obtained by derivating the conditional c.d.f. (see 5.3). Since we have now at hand some estimator $\widehat{F}_Y^{\chi}$ of F_Y^{χ}, it is natural to propose the following estimate:

$$\widehat{f}_Y^{\chi}(\chi, y) = \frac{\partial}{\partial y} \widehat{F}_Y^{\chi}. \tag{5.29}$$

Assuming the differentiability of H, we have

$$\frac{\partial}{\partial y} \widehat{F}_Y^{\chi} = \frac{\sum_{i=1}^n K \left(h^{-1} d(\chi, \boldsymbol{\mathcal{X}}_i) \right) \frac{\partial}{\partial y} H \left(g^{-1} (y - Y_i) \right)}{\sum_{i=1}^n K \left(h^{-1} d(\chi, \boldsymbol{\mathcal{X}}_i) \right)},$$

and this is motivating the following expression for the kernel functional conditional density estimate:

$$\widehat{f}_Y^{\boldsymbol{\chi}}(\chi, y) = \frac{\sum_{i=1}^n K \left(h^{-1} d(\chi, \boldsymbol{\mathcal{X}}_i) \right) \frac{1}{g} H' \left(g^{-1} (y - Y_i) \right)}{\sum_{i=1}^n K \left(h^{-1} d(\chi, \boldsymbol{\mathcal{X}}_i) \right)}.$$

More generally, we can state for any kernel K_0 the following definition:

$$\widehat{f}_Y^{\boldsymbol{\chi}}(\chi, y) = \frac{\sum_{i=1}^n K \left(h^{-1} d(\chi, \boldsymbol{\mathcal{X}}_i) \right) \frac{1}{g} H K_0 \left(g^{-1} (y - Y_i) \right)}{\sum_{i=1}^n K \left(h^{-1} d(\chi, \boldsymbol{\mathcal{X}}_i) \right)}. \tag{5.30}$$

This kind of estimate has been widely studied in the un-functional setting, that is, in the setting when $\boldsymbol{\mathcal{X}}$ is changed into a finite dimensional variable (see the bibliographical discussion at the end of Section 6.4.2). Concerning the parameters involved in the definition of this estimate, let us just note that the roles of those involved in the functional part of the estimate (namely, the roles of K and h) are the same as in the regression setting discussed just before while those involved in the un-functional part (namely, K_0 and g) are acting exactly as K and h, respectively as a weight function and as a smoothing factor.

To end, note that we can easily get the following kernel functional conditional mode estimator of $\theta(\chi)$:

$$\widehat{\theta}(\chi) \;=\; \arg\sup_{y \in S} \widehat{f_Y^{\chi}}(\chi, y). \tag{5.31}$$

6

Some Selected Asymptotics

This chapter states a few asymptotic results, all related to the problem of predicting some real random response variable given some explanatory variable which is allowed to be of infinite dimension. We have arbitrarily decided to organize the presentation around the problems of estimating the three functional predictors defined in Chapter 5: the regression, the conditional median and the conditional mode. Additional results on conditional quantiles, on conditional functional c.d.f. and on conditional density estimation problems will also be given. Even if this field of statistics is quite new, it was impossible to present all the results which are actually available in the literature. The necessary selection was done according to two wishes: providing self-contained and detailed proofs of some key results without hiding the main features of functional problems with too much technicality. This is the reason why some of the results are not presented under the most sophisticated sets of assumptions. A final section will complete these results by discussing the relevant bibliography and by giving a prominent place to the statements of open problems.

6.1 Introduction

The aim of this chapter is to present some asymptotic results linked with nonparametric estimation of the three functional predictors already defined in Section 5.2: conditional expectation, conditional median and conditional mode. The nonparametric predictors to be used are based on kernel smoothing ideas and are those defined before in Section 5.4. All the results are presented in terms of almost complete convergence. This stochastic mode of convergence may appear quite unusual for some people, but it has been selected because it has two important advantages. First it is stronger than almost sure convergence and convergence in probability, and second it is easier to prove than the almost sure consistency itself. To help the reader who is not very familiar with this kind of asymptotics, Appendix A recalls the basic definitions coming with almost complete convergence and shows its links with other usual modes of convergence.

The asymptotic results are divided into two parts: convergence results are first stated in Section 6.2 and the rates of convergence are stated precisely later on Section 6.3. Of course, the treatment of each of the three functional predictors (conditional expectation, conditional median and conditional mode) involves necessarily the study of its associated nonlinear operator: the regression, the conditional c.d.f. and the conditional density function. Therefore, each of these sections is divided in five parts: one for each of the three functional nonparametric predictors, one for the extension of conditional median results to conditional quantiles, and a last one for summarizing and completing different results concerning the estimation of the nonlinear operators associated with the conditional distribution of the process. As it is basically the case in un-functional (i.e., finite dimensional) situations, the statement of consistency results relies on continuity assumptions on the predictor or on the nonlinear operator to be estimated, and naturally the results of Section 6.2 will be stated on continuity-type models like those described in Section 5.3. As in finite dimensional settings, the specification of the rates of convergence relies on additional smoothness conditions and the results of Section 6.3 will be stated under Lipschitz-type conditions under the functional parts of the nonlinear operators to be considered. Rather then looking for the most sophisticated technicalities, we decided to emphasize on results for which the specificities of the infinite dimensionality of the explanatory variables can be highlighted. Each of the results presented is accompanied by a complete proof.

A final Section 6.4 is devoted to the discussion of all the results given in this chapter. This discussion is organized around three ideas. First, we present an up-to-date survey of the quite few bibliography existing in this new setting of nonparametric statistics for functional prediction problems. Second, to emphasize the specificity of the functional feature of the problem, we will show how the results behave in finite dimensional setting. All the results of this chapter lead us to think that large parts of the nonparametric knowledges for finite dimensional statistics could be transplanted to infinite dimensional settings (depending, of course, on suitable functional adaptations). Therefore, to share this hope with the statisticians community and to encourage further investigation in this direction, the third and last (but not least) thread of these comments will be to release several theoretical open problems.

6.2 Almost Complete Convergence

6.2.1 Regression Estimation

We focus on the pointwise almost complete convergence of the functional kernel estimator of the regression $r(\mathcal{X}) = \mathbb{E}(Y|\mathcal{X})$, which was defined in 5.23. Note that we can write

$$Y = r(\boldsymbol{X}) + \varepsilon \text{ with } \mathbb{E}\left(\varepsilon | \boldsymbol{X}\right) = 0. \tag{6.1}$$

Before giving the main asymptotic result, we need to make some assumptions. The first one is about the small ball probabilities of the functional variable $\boldsymbol{X}$, and it means that the probability of observing the f.r.v. around χ (the functional element at which we evaluate the regression operator) is nonnull:

$$\forall \epsilon > 0, \; P\left(\boldsymbol{X} \in B(\chi, \epsilon)\right) = \varphi_{\chi}(\epsilon) > 0. \tag{6.2}$$

It is classical in the multivariate nonparametric setting to assume that the density of the multivariate explanatory variable is strictly positive, and the hypothesis (6.2) is an extension of such a notion (see Sections 13.4 and 13.5 for more discussion). In addition, the parameters involved in the estimator, that is the bandwidth and the kernel function, have to satisfy:

$$\left\{ \begin{array}{c} h \text{ is a positive sequence such that} \\ \lim_{n \to \infty} h = 0 \text{ and } \lim_{n \to \infty} \dfrac{\log n}{n \, \varphi_{\chi}(h)} = 0, \\[2mm] K \text{ is a kernel of type I} \\ or \\ K \text{ is a kernel of type II and (4.6) holds.} \end{array} \right. \tag{6.3}$$

Finally, we will consider a scalar response variable Y such that:

$$\forall m \geq 2, \; \mathbb{E}\left(|Y^m| \,|\, \boldsymbol{X} = \chi\right) < \sigma_m(\chi) < \infty \text{ with } \sigma_m(.) \text{ continuous at } \chi. \tag{6.4}$$

This assumption allows us to deal with unbounded variables. The following theorem gives the pointwise almost complete convergence for the kernel estimator of the nonlinear regression operator.

Theorem 6.1. *Under the continuity-type model (5.11) with the probability condition (6.2), if the estimator verifies (6.3) and if the response variable Y satisfies (6.4), then we have:*

$$\lim_{n \to \infty} \widehat{r}(\chi) = r(\chi), \; a.co. \tag{6.5}$$

Proof. Using the notation introduced in Section 4.1, for $i = 1, \dots, n$, we recall that Δ_i is the quantity defined as

$$\Delta_i = \frac{K\left(h^{-1} d(\chi, \boldsymbol{X}_i)\right)}{\mathbb{E} \, K\left(h^{-1} d(\chi, \boldsymbol{X}_1)\right)}.$$

Note that Lemma 4.3 and Lemma 4.4, together with (6.2), ensure that $\mathbb{E} \, K\left(h^{-1} d(\chi, \boldsymbol{X}_1)\right) > 0$. Let $\widehat{r}_1(\chi)$ and $\widehat{r}_2(\chi)$ be the following quantities:

$$\widehat{r}_1(\chi) = \frac{1}{n} \sum_{i=1}^{n} \Delta_i \qquad (6.6)$$

and

$$\widehat{r}_2(\chi) = \frac{1}{n} \sum_{i=1}^{n} Y_i \Delta_i.$$

We have clearly that $\widehat{r}(\chi) = \widehat{r}_2(\chi)/\widehat{r}_1(\chi)$. Our proof is based on the decomposition

$$\widehat{r}(\chi) - r(\chi) = \frac{1}{\widehat{r}_1(\chi)} \left\{ (\widehat{r}_2(\chi) - \mathbb{E}\widehat{r}_2(\chi)) - (r(\chi) - \mathbb{E}\widehat{r}_2(\chi)) \right\}$$
$$- \frac{r(\chi)}{\widehat{r}_1(\chi)} \left\{ \widehat{r}_1(\chi) - 1 \right\}. \qquad (6.7)$$

The numerators in this decomposition will be treated directly by using Lemma 6.2 and Lemma 6.3 below, while the denominators are treated directly by using again part ii) of Lemma 6.3 together with part i) of Proposition A.6. Finally the proof of Theorem 6.1 is finished, at least as long as both of the following lemmas will be checked. □

Lemma 6.2. *Under (5.11) and (6.3) we have:*

$$\lim_{n \to \infty} \mathbb{E}\widehat{r}_2(\chi) = r(\chi).$$

Proof. The model (6.1) allows us directly to write:

$$r(\chi) - \mathbb{E}\widehat{r}_2(\chi) = r(\chi) - \mathbb{E}(Y_1 \Delta_1),$$
$$= r(\chi) - \mathbb{E}\left(\mathbb{E}\left(Y_1 \Delta_1 | \boldsymbol{X}_1 \right) \right),$$
$$= r(\chi) - \mathbb{E}\left(r(\boldsymbol{X}_1) \Delta_1 \right),$$
$$= \mathbb{E}\left((r(\chi) - r(\boldsymbol{X}_1)) \Delta_1 \right).$$

Because the support of the kernel function K is $[0, 1]$, we have:

$$|r(\chi) - r(\boldsymbol{X}_1)| \Delta_1 \leq \sup_{\chi' \in B(\chi, h)} |r(\chi) - r(\chi')| \Delta_1,$$

and the continuity assumption on r allows to get the claimed result. □

Lemma 6.3. *We have:*

i) Under assumptions (6.2), (6.3) and (6.4), we have:

$$\widehat{r}_2(\chi) - \mathbb{E}\widehat{r}_2(\chi) = O_{a.co.}\left(\sqrt{\frac{\log n}{n\, \varphi_\chi(h)}} \right),$$

ii) Under assumptions (6.2) and (6.3), it holds:

$$\widehat{r}_1(\chi) - 1 = O_{a.co.} \left(\sqrt{\frac{\log n}{n\,\varphi_\chi(h)}} \right).$$

Proof. i) We denote, for $i = 1, \ldots, n$, $K_i = K\left(h^{-1} d(\chi, \boldsymbol{X}_i)\right)$. The demonstration of this result is based on the utilization of a Bernstein-type exponential inequality. Indeed,

$$P\left(|\widehat{r}_2(\chi) - \mathbb{E}\widehat{r}_2(\chi)| > \epsilon\right) = P\left(\frac{1}{n}\left|\sum_{i=1}^{n}(Y_i\Delta_i - \mathbb{E}(Y_i\Delta_i))\right| > \epsilon\right),$$

and we have to show that it exists $\epsilon_0 > 0$ such that:

$$\sum_{n\in\mathbb{N}^*} P\left(\frac{1}{n}\left|\sum_{i=1}^{n}(Y_i\Delta_i - \mathbb{E}(Y_i\Delta_i))\right| > \epsilon_0\sqrt{\frac{\log n}{n\,\varphi_\chi(h)}}\right) < \infty.$$

So, we apply the exponential inequality given by Corollary A.8-ii in Appendix A with $Z_i = Y_i\Delta_i - \mathbb{E}Y_1\Delta_1$. To do that, we first have to show that:

$$\exists C > 0, \ \forall m = 2, 3, \ldots, \ \left|\mathbb{E}\left(Y_1\Delta_1 - \mathbb{E}Y_1\Delta_1\right)^m\right| \leq C\,\varphi_\chi(h)^{-m+1}. \quad (6.8)$$

- First, we prove that for $m \geq 2$:

$$\mathbb{E}|Y_1|^m\Delta_1^m = O\left(\varphi_\chi(h)^{-m+1}\right). \quad (6.9)$$

For that, we write:

$$\mathbb{E}|Y_1|^m\Delta_1^m = \frac{1}{(\mathbb{E}K_1)^m}\left\{\mathbb{E}\,|Y_1|^m\,K_1^m\right\},$$

$$= \frac{1}{(\mathbb{E}K_1)^m}\left\{\mathbb{E}\left(\mathbb{E}(|Y_1|^m|\boldsymbol{X})K_1^m\right)\right\},$$

$$= \frac{1}{(\mathbb{E}K_1)^m}\left\{\mathbb{E}\sigma_m(\boldsymbol{X})K_1^m\right\},$$

$$= \frac{1}{(\mathbb{E}K_1)^m}\left\{\mathbb{E}\left((\sigma_m(\boldsymbol{X}) - \sigma_m(\chi))K_1^m\right) + \sigma_m(\chi)\,\mathbb{E}\,K_1^m\right\},$$

which implies that:

$$|\mathbb{E}|Y_1|^m\Delta_1^m| \leq \mathbb{E}|\sigma_m(\boldsymbol{X}) - \sigma_m(\chi)|\Delta_1^m + \sigma_m(\chi)\,\mathbb{E}\Delta_1^m,$$

$$\leq \left(\sup_{\chi'\in B(\chi,h)}|\sigma_m(\chi') - \sigma_m(\chi)|\right)\mathbb{E}\Delta_1^m + \sigma_m(\chi)\mathbb{E}\Delta_1^m.$$

Because $0 < \int K^m < \infty$, if K is of type I (resp. II) then $K^m / \int K^m$ is also of type I (resp. II). So, by applying Lemma 4.3 and Lemma 4.4 we get:

$$C_1 \, \varphi_\chi(h) \leq \mathbb{E} \, K_1^m \leq C_2 \, \varphi_\chi(h). \qquad (6.10)$$

Using (6.10) and Lemma 4.3 or Lemma 4.4, we can write that for $m = 2, 3, \ldots$:

$$\frac{C_1}{\varphi_\chi(h)^{m-1}} \leq \mathbb{E}\Delta_1^m \leq \frac{C_2}{\varphi_\chi(h)^{m-1}}, \qquad (6.11)$$

which implies that

$$\mathbb{E} \, |Y_1|^m \, \Delta_1^m = O\left(\varphi_\chi(h)^{-m+1}\right).$$

- Moreover, we have:

$$(Y_1\Delta_1 - \mathbb{E}Y_1\Delta_1)^m = \sum_{k=0}^m c_{k,m}(Y_1\Delta_1)^k (\mathbb{E}Y_1\Delta_1)^{m-k}(-1)^{m-k},$$

where $c_{k,m} = m!/(k!(m-k)!)$, which implies that

$$\mathbb{E} \, |Y_1\Delta_1 - \mathbb{E}Y_1\Delta_1|^m \leq C \sum_{k=0}^m c_{k,m}\mathbb{E} \, |Y_1\Delta_1|^k \, |r(\chi)|^{m-k}$$

$$\leq C \max_{k=0,1,\ldots,m} \mathbb{E} \, |Y_1\Delta_1|^k \,|$$

$$\leq C \max_{k=0,1,\ldots,m} \varphi_\chi(h)^{-k+1}$$

the last inequality is using (6.9) for $k \geq 2$ while for $k = 1$ we can show that $E|Y_1|\Delta_1 = O(1)$ just by following the same steps as those of the proof of Lemma 6.2. Because $\varphi_\chi(h)$ tends to zero with n, it becomes that

$$\mathbb{E} \, |Y_1\Delta_1 - \mathbb{E}Y_1\Delta_1|^m = O\left((\varphi_\chi(h))^{-m+1}\right).$$

The last point consists of applying Corollary A.8-ii with $a^2 = \varphi_\chi(h)^{-1}$. Then, we have $u_n = (a^2 \log n)/n = \log n/(n \, \varphi_\chi(h))$ and it is clear that u_n tends to zero with n by using Hypothesis (6.3). This ends the proof of Lemma 6.3-i.

ii) This result can be derived directly from (i) by taking $Y_i = 1$.
The proof of Lemma 6.3 is now finished. $\square$

6.2.2 Conditional Median Estimation

We will now give the same kind of consistency result, but for the functional conditional median $m(\chi)$ and its kernel estimate $\widehat{m}(\chi)$, which are respectively defined in (5.5) and (5.27). Recall that the conditional c.d.f. estimate $\widehat{F}_Y^{\boldsymbol{x}}$ was defined in (5.26). Because the asymptotic results will be given at a fixed $\chi \in E$, we simplify some notation as follows:

$$\forall \chi \in E, \ F_Y^\chi(.) \overset{\text{def}}{=} F_Y^{\boldsymbol{x}}(\chi, .) \text{ and } \widehat{F}_Y^\chi(.) \overset{\text{def}}{=} \widehat{F}_Y^{\boldsymbol{x}}(\chi, .). \tag{6.12}$$

For convenience, we will use also the notation:

$$\forall \chi \in E, \ F_Y^{\boldsymbol{x}}(.) \overset{\text{def}}{=} F_Y^{\boldsymbol{x}}(\boldsymbol{x}, .). \tag{6.13}$$

Note that the functional part of $\widehat{F}_Y^{\boldsymbol{x}}$ is the same as in regression setting. Therefore, it is natural to expect the asumptions necessary to deal with this functional part to be the same as in Section 6.2.1. Concerning the scalar part of $\widehat{F}_Y^{\boldsymbol{x}}$ the following restrictions on the kernel function $K_0 = H'$ and its associated bandwidth g are introduced:

$$\begin{cases} g \text{ is a positive sequence such that } \lim_{n\to\infty} g = 0, \\ \quad K_0 \text{ is of type } 0. \end{cases} \tag{6.14}$$

Note this is a quite weak condition on the kernel. It insures, according to Definition 4.2, that the function H is continuous and strictly increasing over the set $\{u, 0 < K(u) < 1\}$. This has the main advantage that the kernel conditional median estimate can be defined to be the unique solution of the equation

$$\widehat{m}(\chi) = \widehat{F}_Y^{\chi -1}(1/2). \tag{6.15}$$

Theorem 6.4. *Under the continuity-type model defined by (5.17) and (6.2), and if the kernel estimate satisfies (6.3) and (6.14), we have*

$$\lim_{n\to\infty} \widehat{m}(\chi) = m(\chi), \quad a.co. \tag{6.16}$$

Proof. The condition (6.14) insures that the estimated conditional c.d.f. $\widehat{F}_Y^\chi(.)$ is continuous and strictly increasing. So, the function $\widehat{F}_Y^{\chi -1}(.)$ exists and is continuous. The continuity property of $\widehat{F}_Y^{\chi -1}(.)$ at point $\widehat{F}_Y^\chi(m(\chi))$ can be written as:

$$\forall \epsilon > 0, \ \exists \delta(\epsilon) > 0, \ \forall y, \ |\widehat{F}_Y^\chi(y) - \widehat{F}_Y^\chi(m(\chi))| \leq \delta(\epsilon) \Rightarrow |y - m(\chi)| \leq \epsilon.$$

In the special case when $y = \widehat{m}(\chi)$, we have

$$\forall \epsilon > 0, \ \exists \delta(\epsilon) > 0, \ |\widehat{F}_Y^\chi(\widehat{m}(\chi)) - \widehat{F}_Y^\chi(m(\chi))| \leq \delta(\epsilon) \Rightarrow |\widehat{m}(\chi) - m(\chi)| \leq \epsilon,$$

in such a way that we arrive finally at:

$$\forall \epsilon > 0, \ \exists \delta(\epsilon) > 0, \ P(|\widehat{m}(\chi) - m(\chi)| > \epsilon)$$
$$\leq P(|\widehat{F}_Y^\chi(\widehat{m}(\chi)) - \widehat{F}_Y^\chi(m(\chi))| > \delta(\epsilon))$$
$$= P(|F_Y^\chi(m(\chi)) - \widehat{F}_Y^\chi(m(\chi))| > \delta(\epsilon)),$$

the last inequality following from the simple observation that

$$F_Y^\chi(m(\chi)) = \widehat{F}_Y^\chi(\widehat{m}(\chi)) = 1/2.$$

The pointwise almost complete convergence of the kernel conditional c.d.f. estimate $\widehat{F}_Y^\chi$ (see Lemma 6.5 below) leads directly to

$$\forall \epsilon > 0, \ \sum_{n=1}^\infty P(|\widehat{m}(\chi) - m(\chi)| > \epsilon) < \infty,$$

and the claimed consistency result (6.16) is now checked. □

Lemma 6.5. *Under the conditions of Theorem 6.4, we have for any fixed real point y:*

$$\lim_{n \to \infty} \widehat{F}_Y^\chi(y) = F_Y^\chi(y), \ a.co. \tag{6.17}$$

Proof. When necessary, the same notation as that introduced in Theorem 6.1 and its proof will be used.

• We can write

$$\widehat{F}_Y^\chi(y) - F_Y^\chi(y) = \frac{\{(\widehat{r}_3(\chi,y) - \mathbb{E}\widehat{r}_3(\chi,y)) - (F_Y^\chi(y) - \mathbb{E}\widehat{r}_3(\chi,y))\}}{\widehat{r}_1(\chi)}$$
$$- \frac{F_Y^\chi(y)}{\widehat{r}_1(\chi)} \{\widehat{r}_1(\chi) - 1\}, \tag{6.18}$$

where $\widehat{r}_1$ is defined by (6.6), and where

$$\widehat{r}_3(\chi,y) = \widehat{r}_1(\chi)\widehat{F}_Y^\chi(y) = \frac{1}{n} \sum_{i=1}^n \Delta_i \Gamma_i(y),$$

with

$$\Gamma_i(y) = H\left(g^{-1}(y - Y_i)\right).$$

Look at the right-hand side of (6.18). Note first that the denominators are directly treated by using Lemma 6.3-ii together with Proposition A.6-i. Note also that the last term is treated by using Lemma 6.3-ii. Finally the claimed result (6.17) will be proved as soon as it is shown that:

$$\lim_{n \to \infty} \mathbb{E}\widehat{r}_3(\chi,y) = F_Y^\chi(y), \tag{6.19}$$

and

$$\lim_{n \to \infty} \widehat{r}_3(\chi,y) - \mathbb{E}\widehat{r}_3(\chi,y) = 0, \ a.co. \tag{6.20}$$

- About the bias term, the fact that $\mathbb{E}\widehat{r}_1(\chi) = 1$ and the fact that K is compactly supported lead directly to

$$
\begin{aligned}
\mathbb{E}\widehat{r}_3(\chi, y) - F_Y^X(y) &= \mathbb{E}\Delta_1\Gamma_1(y) - F_Y^X(y) \\
&= \mathbb{E}(\Delta_1(\mathbb{E}(\Gamma_1(y)|\boldsymbol{X}) - F_Y^X(y))) \\
&= \mathbb{E}(\Delta_1 1_{B(\chi,h)}(\boldsymbol{X})(\mathbb{E}(\Gamma_1(y)|\boldsymbol{X}) - F_Y^X(y))). \quad (6.21)
\end{aligned}
$$

This last expectation can be easily computed by means of the Fubini theorem and by using the fact that $H' = K_0$:

$$
\begin{aligned}
\mathbb{E}(\Gamma_1(y)|\boldsymbol{X}) &= \int_{\mathbb{R}} H(g^{-1}(y-u))dP(u|\boldsymbol{X}) \\
&= \int_{\mathbb{R}} \int_{-\infty}^{\frac{y-u}{g}} K_0(v)dv\, dP(u|\boldsymbol{X}) \\
&= \int_{\mathbb{R}} \int_{\mathbb{R}} K_0(v)1_{[v,+\infty]}(g^{-1}(y-u))dv\, dP(u|\boldsymbol{X}) \\
&= \int_{\mathbb{R}} \int_{\mathbb{R}} K_0(v)1_{[v,+\infty]}(g^{-1}(y-u))dP(u|\boldsymbol{X})\, dv \\
&= \int_{\mathbb{R}} K_0(v)F_Y^{\boldsymbol{X}}(y-vg)dv. \quad (6.22)
\end{aligned}
$$

Because K_0 integrates up to 1, we get :

$$
\mathbb{E}(\Gamma_1(y)|\boldsymbol{X}) - F_Y^X(y) = \int_{\mathbb{R}} K_0(v)(F_Y^{\boldsymbol{X}}(y-vg) - F_Y^X(y))dv. \quad (6.23)
$$

Moreover, we have:

$$
|F_Y^{\boldsymbol{X}}(y-vg) - F_Y^X(y)| \leq |F_Y^{\boldsymbol{X}}(y-vg) - F_Y^X(y-vg)| + |F_Y^X(y-vg) - F_Y^X(y)|.
$$

Because K_0 is supported on $[-1,+1]$ and because g and h tend to zero, the continuity property of $F_Y^{\boldsymbol{X}}$ allows us to write that

$$
\lim_{n\to\infty} \sup_{v\in[-1,+1]} 1_{B(\chi,h)}(\boldsymbol{X})|F_Y^{\boldsymbol{X}}(y-vg) - F_Y^X(y-vg)| = 0,
$$

and

$$
\lim_{n\to\infty} \sup_{v\in[-1,+1]} |F_Y^X(y-vg) - F_Y^X(y)| = 0,
$$

in such a way that we arrive finally at

$$
\lim_{n\to\infty} 1_{B(\chi,h)}(\boldsymbol{X})|\mathbb{E}(\Gamma_1(y)|\boldsymbol{X}) - F_Y^X(y))| = 0. \quad (6.24)
$$

This last result, together with (6.21) and the fact that $\mathbb{E}\Delta_1 = 1$ is enough to prove (6.19).

- It just remains now to prove (6.20). For that, we have to decompose the dispersion term as a sum of zero mean i.i.d. r.r.v. as follows:

$$\widehat{r}_3(\chi, y) - \mathbb{E}\widehat{r}_3(\chi, y) = \frac{1}{n} \sum_{i=1}^{n} (T_i - \mathbb{E}T_i), \qquad (6.25)$$

where

$$T_i = \Delta_i \Gamma_i(y).$$

Using either Lemma 4.3 or Lemma 4.4 according to the fact that K is of type I or II, and because $\Gamma_i(y)$ is bounded, we have:

$$T_i \leq C/\varphi_\chi(h). \qquad (6.26)$$

On the other hand, by applying the result (6.9) with $m = 2$, we have

$$\mathbb{E}T_i^2 \leq C \, \mathbb{E}\Delta_i^2 \leq C/\varphi_\chi(h). \qquad (6.27)$$

Because the variables T_i are bounded, the results (6.26) and (6.27) are enough to treat the term T_i. Precisely, we are now in position to apply the Bernstein-type exponential inequality given by Corollary A.9-ii (see Appendix A), and we get:

$$\widehat{r}_3(\chi, y) - \mathbb{E}\widehat{r}_3(\chi, y) = O_{a.co.}\left(\frac{\log n}{n\varphi_\chi(h)}\right)^{\frac{1}{2}}. \qquad (6.28)$$

which is a stronger result than the claimed one (6.20).

This proof is now finished. □

6.2.3 Conditional Mode Estimation

We will now give the same kind of consistency result, but for the conditional mode $\theta(\chi)$ and its kernel estimate $\widehat{\theta}(\chi)$, which are respectively defined in (5.20) and (5.31). Because we focus on pointwise convergence at a fixed χ lying in E, we consider the following simplified notation:

$$\forall \chi \in E, \; f_Y^\chi(.) \stackrel{\text{def}}{=} f_Y^\chi(\chi, .) \text{ and } \widehat{f}_Y^\chi(.) \stackrel{\text{def}}{=} \widehat{f}_Y^\chi(\chi, .). \qquad (6.29)$$

For convenience, we will use also the notation:

$$\forall \chi \in E, \; f_Y^{\boldsymbol{\chi}}(.) \stackrel{\text{def}}{=} f_Y^{\boldsymbol{\chi}}(\boldsymbol{\mathcal{X}}, .). \qquad (6.30)$$

The assumptions needed for the kernel conditional density estimate $\widehat{f}_Y^{\boldsymbol{\chi}}$ defined by (5.29) are closed to those introduced in Section 6.2.2 for estimating the conditional c.d.f. $\widehat{F}_Y^\chi$. The following unrestrictive assumptions have to be added:

$$\begin{cases} \exists C < \infty, \ \forall (x, x') \in \mathbb{R} \times \mathbb{R}, \ |K_0(x) - K_0(x')| \leq C|x - x'|, \\ \\ \lim_{n \to \infty} \dfrac{\log n}{n \, g \, \varphi_\chi(h)} = 0 \text{ and } \exists \zeta > 0, \ \lim_{n \to \infty} g n^\zeta = \infty, \end{cases} \qquad (6.31)$$

Theorem 6.6. *Under the continuity-type model defined by (5.21) and (6.2), and if the kernel estimate satisfies (6.3), (6.14) and (6.31), we have*

$$\lim_{n \to \infty} \widehat{\theta}(\chi) = \theta(\chi), \quad a.co. \qquad (6.32)$$

Proof. The condition (5.21) insures that the true conditional density $f_Y^\chi(.)$ is continuous and strictly increasing on $(\theta(\chi) - \xi, \theta(\chi))$. So, the function $f_Y^{\chi^{-1}}(.)$ exists and is continuous. The continuity property of $f_Y^{\chi^{-1}}(.)$ at point $f_Y^\chi(\theta(\chi))$ can be written for any $\epsilon > 0$ as:

$$\exists \delta_1(\epsilon) > 0, \ \forall y \in (\theta(\chi) - \xi, \theta(\chi)), \ |f_Y^\chi(y) - f_Y^\chi(\theta(\chi))| \leq \delta_1(\epsilon) \Rightarrow |y - \theta(\chi)| \leq \epsilon.$$

Because $\widehat{f}_Y^\chi(.)$ is continuous and strictly decreasing on $(\theta(\chi), \theta(\chi) + \chi)$, the same kind of argument can be invoked to arrive at:

$$\exists \delta_2(\epsilon) > 0, \ \forall y \in (\theta(\chi), \theta(\chi) + \xi), \ |f_Y^\chi(y) - f_Y^\chi(\theta(\chi))| \leq \delta_2(\epsilon) \Rightarrow |y - \theta(\chi)| \leq \epsilon.$$

By combining both results, we arrive at:

$$\exists \delta(\epsilon) > 0, \ \forall y \in (\theta(\chi) - \xi, \theta(\chi) + \xi), \ |f_Y^\chi(y) - f_Y^\chi(\theta(\chi))| \leq \delta(\epsilon) \Rightarrow |y - \theta(\chi)| \leq \epsilon.$$

Because by construction $\widehat{\theta}(\chi) \in (\theta(\chi) - \xi, \theta(\chi) + \xi)$, we have:

$$\exists \delta(\epsilon) > 0, \ |f_Y^\chi(\widehat{\theta}(\chi)) - f_Y^\chi(\theta(\chi))| \leq \delta(\epsilon) \Rightarrow |y - \theta(\chi)| \leq \epsilon,$$

so that we arrive finally at:

$$\exists \delta(\epsilon) > 0, \ P(|\widehat{\theta}(\chi) - \theta(\chi)| > \epsilon) \leq P(|f_Y^\chi(\widehat{\theta}(\chi)) - f_Y^\chi(\theta(\chi))| > \delta(\epsilon)).$$

On the other hand, it follows directly from the definitions of $\theta(\chi)$ and $\widehat{\theta}(\chi)$ that:

$$\begin{aligned} |f_Y^\chi(\theta(\chi)) - f_Y^\chi(\widehat{\theta}(\chi))| &= f_Y^\chi(\theta(\chi)) - f_Y^\chi(\widehat{\theta}(\chi)) \\ &= \left(f_Y^\chi(\theta(\chi)) - \widehat{f}_Y^\chi(\theta(\chi)) \right) + \left(\widehat{f}_Y^\chi(\theta(\chi)) - f_Y^\chi(\widehat{\theta}(\chi)) \right) \\ &\leq \left(f_Y^\chi(\theta(\chi)) - \widehat{f}_Y^\chi(\theta(\chi)) \right) + \left(\widehat{f}_Y^\chi(\widehat{\theta}(\chi)) - f_Y^\chi(\widehat{\theta}(\chi)) \right) \\ &\leq 2 \sup_{y \in (\theta(\chi) - \xi, \theta(\chi) + \xi)} |f_Y^\chi(y) - \widehat{f}_Y^\chi(y)|. \end{aligned} \qquad (6.33)$$

The uniform complete convergence of the kernel conditional density estimate over the compact set $[\theta(\chi) - \xi, \theta(\chi) + \xi]$ (see Lemma 6.7 below) can be used, leading directly from both previous inequalities to:

$$\forall \epsilon > 0, \sum_{n=1}^{\infty} P\left(|\widehat{\theta}(\chi) - \theta(\chi)| > \epsilon\right) < \infty.$$

Finally, the claimed consistency result (6.32) will be proved as long as the following lemma could be checked. $\square$

Lemma 6.7. *Under the conditions of Theorem 6.6, we have for any compact subset $S \subset \mathbb{R}$:*

$$\lim_{n \to \infty} \sup_{y \in S} |f_Y^{\chi}(y) - \widehat{f}_Y^{\chi}(y)| = 0, \ a.co. \tag{6.34}$$

Proof. The proof is based on the following decomposition

$$\widehat{f}_Y^{\chi}(y) - f_Y^{\chi}(y) = \frac{\{(\widehat{r}_4(\chi, y) - \mathbb{E}\widehat{r}_4(\chi, y)) - (f_Y^{\chi}(y) - \mathbb{E}\widehat{r}_4(\chi, y))\}}{\widehat{r}_1(\chi)}$$

$$- \frac{f_Y^{\chi}(y)}{\widehat{r}_1(\chi)} \{\widehat{r}_1(\chi) - 1\}, \tag{6.35}$$

where $\widehat{r}_1$ is defined by (6.6), and where

$$\widehat{r}_4(\chi, y) = \widehat{r}_1(\chi)\widehat{f}_Y^{\chi}(y) = \frac{1}{n} \sum_{i=1}^{n} \Delta_i \Omega_i(y),$$

with

$$\Omega_i(y) = g^{-1} K_0\left(g^{-1}(y - Y_i)\right).$$

Look at the right-hand side of (6.35). Note first that the denominators are directly treated by using Lemma 6.3-ii together with Proposition A.6-i. Note also that the last term is treated by using also Lemma 6.3-ii and that $f_Y^{\chi}(.)$ is uniformly bounded over $y \in S$ (since it is continuous on the compact set S). Therefore, the result (6.34) will be a direct consequence of both following ones:

$$\lim_{n \to \infty} \frac{1}{\widehat{r}_1(\chi)} \sup_{y \in S} |\mathbb{E}\widehat{r}_4(\chi, y) - f_Y^{\chi}(y)| = 0, \ a.co., \tag{6.36}$$

and

$$\lim_{n \to \infty} \frac{1}{\widehat{r}_1(\chi)} \sup_{y \in S} |\widehat{r}_4(\chi, y) - \mathbb{E}\widehat{r}_4(\chi, y)| = 0, \ a.co. \tag{6.37}$$

- Let us show the result (6.36). Because $\mathbb{E}\Delta_1 = 1$ and because K_0 is integrating up to one, we have, after integration by substitution:

$$
\begin{aligned}
\mathbb{E}\widehat{r_4}(\chi, y) - f_Y^\chi(y) &= \mathbb{E}\Delta_1 \Omega_1(y) - f_Y^\chi(y) \\
&= \mathbb{E}\left(\Delta_1 (\mathbb{E}(\Omega_1(y)|\mathcal{X}) - f_Y^\chi(y))\right) \\
&= \mathbb{E}\left(\Delta_1 (\int_{\mathbb{R}} g^{-1} K_0 \left(g^{-1}(y-u)\right) f_Y^{\mathcal{X}}(u) du - f_Y^\chi(y))\right) \\
&= \mathbb{E}\left(\Delta_1 \int_{\mathbb{R}} g^{-1} K_0 \left(g^{-1}(y-u)\right) (f_Y^{\mathcal{X}}(u) - f_Y^\chi(y)) du\right) \\
&= \mathbb{E}\left(\Delta_1 \int_{\mathbb{R}} K_0(v)(f_Y^{\mathcal{X}}(y-vg) - f_Y^\chi(y)) dv\right) \\
&= \mathbb{E}\left(1_{B(\chi,h)}(\mathcal{X}) \Delta_1 \int_{\mathbb{R}} K_0(v)\right. \\
&\qquad\qquad \left. \times (f_Y^{\mathcal{X}}(y-vg) - f_Y^\chi(y)) dv\right). \quad (6.38)
\end{aligned}
$$

In addition, we can write:

$$
\left|f_Y^{\mathcal{X}}(y-vg) - f_Y^\chi(y)\right| \le \left|f_Y^{\mathcal{X}}(y-vg) - f_Y^\chi(y-vg)\right| + \left|f_Y^\chi(y-vg) - f_Y^\chi(y)\right|.
$$

Because S is compact, the function f_Y^χ is uniformly continuous over $y \in S$. Since in addition K_0 is compactly supported (because it is of type 0), and because the continuity-type model (5.21), we have

$$
\lim_{n\to\infty} \sup_{v\in[-1,+1]} \sup_{y\in S} 1_{B(\chi,h)}(\mathcal{X}) |f_Y^{\mathcal{X}}(y-vg) - f_Y^\chi(y-vg)| = 0,
$$

and

$$
\lim_{n\to\infty} \sup_{v\in[-1,+1]} \sup_{y\in S} |f_Y^\chi(y-vg) - f_Y^\chi(y)| = 0,
$$

which implies that

$$
\lim_{n\to\infty} \sup_{v\in[-1,+1]} \sup_{y\in S} 1_{B(\chi,h)}(\mathcal{X}) |f_Y^{\mathcal{X}}(y-vg) - f_Y^\chi(y)| = 0. \quad (6.39)
$$

By combining (6.38) and (6.39), together with the positiveness of Δ_1 and K_0, we get:

$$
\sup_{y\in S} |\mathbb{E}\widehat{r_4}(\chi, y) - f_Y^\chi(y)| = = o(1). \quad (6.40)
$$

This, combined with Lemma 6.3-ii and with Proposition A.6-i, is enough to prove (6.36) and to finish the treatment of this bias term.

- It remains to check that (6.37) is true. Using the compactness of S, we can write that $S \subset \bigcup_{k=1}^{z_n} S_k$ where $S_k = (t_k - l_n, \ t_k + l_n)$ and where l_n and z_n can be chosen such that:

$$
l_n = C z_n^{-1} \sim C n^{-2\varsigma}. \quad (6.41)
$$

Taking $t_y = \arg\min_{t\in\{t_1,\dots,t_{z_n}\}} |y - t|$, we have

$$\frac{1}{\widehat{r}_1(\chi)} \sup_{y \in S} |\widehat{r}_4(\chi, y) - \mathbb{E}\widehat{r}_4(\chi, y)| = A_1 + A_2 + A_3, \qquad (6.42)$$

where

$$A_1 = \frac{1}{\widehat{r}_1(\chi)} \sup_{y \in S} |\widehat{r}_4(\chi, y) - \widehat{r}_4(\chi, t_y)|,$$

$$A_2 = \frac{1}{\widehat{r}_1(\chi)} \sup_{y \in S} |\widehat{r}_4(\chi, t_y) - \mathbb{E}\widehat{r}_4(\chi, t_y)|,$$

$$A_3 = \frac{1}{\widehat{r}_1(\chi)} \sup_{y \in S} |\mathbb{E}\widehat{r}_4(\chi, t_y) - \mathbb{E}\widehat{r}_4(\chi, y)|.$$

The Hölder continuity condition (6.31) allows us directly to write:

$$
\begin{aligned}
|\widehat{r}_4(\chi, y) - \widehat{r}_4(\chi, t_y)| &= \frac{1}{n} \sum_{i=1}^{n} \Delta_i |\Omega_i(y) - \Omega_i(t_y)| \\
&\leq \frac{1}{ng} \sum_{i=1}^{n} \Delta_i |K_0(g^{-1}(y - Y_i)) - K_0(g^{-1}(t_y - Y_i))| \\
&\leq \frac{C}{ng} \sum_{i=1}^{n} \Delta_i \frac{|y - t_y|}{g} \\
&\leq C\widehat{r}_1(\chi) l_n g^{-2}.
\end{aligned}
$$

Using (6.41), it holds that $A_1 \leq C/(g n^\varsigma)^2$ and (6.31) implies that:

$$\lim_{n \to \infty} A_1 = 0. \qquad (6.43)$$

Following similar arguments, we can write:

$$A_3 \leq \frac{C}{\widehat{r}_1(\chi)(g n^\varsigma)^2}, \qquad (6.44)$$

and according to Lemma 6.3-ii and Proposition A.6-i, we get:

$$\lim_{n \to \infty} A_3 = 0, \ a.co. \qquad (6.45)$$

Looking now at the term A_2 we can write for any $\epsilon > 0$:

$$
\begin{aligned}
P(\sup_{y \in S} |\widehat{r}_4(\chi, t_y) - \mathbb{E}\widehat{r}_4(\chi, t_y)| > \epsilon) &= \\
&= P(\max_{j=1\ldots,z_n} |\widehat{r}_4(\chi, t_j) - \mathbb{E}\widehat{r}_4(\chi, t_j)| > \epsilon) \\
&\leq z_n \max_{j=1\ldots,z_n} P(|\widehat{r}_4(\chi, t_j) - \mathbb{E}\widehat{r}_4(\chi, t_j)| > \epsilon) \\
&\leq z_n \max_{j=1\ldots,z_n} P\left(\left|\frac{1}{n} \sum_{i=1}^{n} (U_i - \mathbb{E}U_i)\right| > \epsilon\right), \quad (6.46)
\end{aligned}
$$

where
$$U_i = \Delta_i \Omega_i(t_j).$$

Using either Lemma 4.3 or Lemma 4.4 according to the fact that K is of type I or II, and because $\Omega_i(y) \leq C/g$, we have:

$$|U_i| \leq C/(g\varphi_\chi(h)). \tag{6.47}$$

On the other hand, we have, after integrating by substitution, and using (6.39):

$$\begin{aligned}
\mathbb{E}U_i^2 &= \mathbb{E}\left(\Delta_i^2 \mathbb{E}(\Omega(t_j)^2|\mathcal{X} = \chi)\right) \\
&= \mathbb{E}\left(\Delta_i^2 (\int_{\mathbb{R}} g^{-2} K_0^2 \left(g^{-1}(t_j - u)\right) f_Y^\chi(u)du\right) \\
&= \mathbb{E}\left(\Delta_i^2 g^{-1} \int_{\mathbb{R}} K_0^2(z) f_Y^\chi(t_j + zg)dz\right) \\
&\leq C\mathbb{E}\left(\Delta_i^2 g^{-1} \int_{\mathbb{R}} K_0^2(z)dz f_Y^\chi(t_j)\right).
\end{aligned}$$

Because f_Y^χ is bounded (since it is continuous over the compact set S) and by applying the result (6.9) with $m = 2$, one get:

$$\mathbb{E}U_i^2 \leq C/(g\varphi_\chi(h)). \tag{6.48}$$

Because the variables U_i are bounded, one is therefore in position to apply the Bernstein-type exponential inequality given by Corollary A.9-i (seeAppendix A). This inequality together with (6.46), (6.47) and (6.48) gives directly:

$$P(\sup_{y \in S} |\widehat{r}_4(\chi, t_y) - \mathbb{E}\widehat{r}_4(\chi, t_y)| > \epsilon) \leq z_n \exp\{-Cn\epsilon^2 g\varphi_\chi(h)\}.$$

By using (6.41), one gets:

$$P(\sup_{y \in S} |\widehat{r}_4(\chi, t_y) - \mathbb{E}\widehat{r}_4(\chi, t_y)| > \epsilon) \leq Cn^{2\varsigma} \exp\{-C\,n\,\epsilon^2\,g\,\varphi_\chi(h)\}.$$

Because $\log n/(n\,g\,\varphi_\chi(h))$ tends to zero, one gets directly:

$$\forall \epsilon > 0, \ \sum_{n=1}^{\infty} P(\sup_{y \in S} |\widehat{r}_4(\chi, t_y) - \mathbb{E}\widehat{r}_4(\chi, t_y)| > \epsilon) < \infty. \tag{6.49}$$

The denominator of A_2 is treated directly by using again Lemma 6.3-ii together with Proposition A.6-i. This is enough to get

$$\lim_{n \to \infty} A_2 = 0, \ a.co. \tag{6.50}$$

Finally, the claimed result (6.37) follows from (6.42), (6.43), (6.45) and (6.50).

The combination of (6.35), (6.36) and (6.37) allows us to finish the proof of this lemma. $\square$

6.2.4 Conditional Quantile Estimation

This section is devoted to the generalization of the results given in Section 6.2.2 to the estimation of conditional functional quantiles. The conditional quantile of order α, denoted by $t_\alpha(\chi)$, and its kernel estimate $\widehat{t}_\alpha(\chi)$ are respectively defined by (5.9) and (5.28). As pointed out in Section 6.2.2, under the condition (6.14), the kernel conditional quantile estimate can be defined to be the unique solution of the equation

$$\widehat{t}_\alpha(\chi) = \widehat{F}_Y^{\chi-1}(\alpha). \tag{6.51}$$

Theorem 6.8. *Let $\alpha \in (0,1)$. Under the continuity-type model defined by (5.17) and (6.2), and if the kernel estimate satisfies (6.3) and (6.14), we have*

$$\lim_{n\to\infty} \widehat{t}_\alpha(\chi) = t_\alpha(\chi), \quad a.co. \tag{6.52}$$

Proof. This proof being closely related to the proof of Theorem 6.16, it will be more briefly presented. Let $\epsilon > 0$ be fixed. The continuity property of $\widehat{F}_Y^{\chi-1}(.)$ at point $\widehat{F}_Y^\chi(t_\alpha(\chi))$ can be written as:

$$\exists \delta(\epsilon) > 0, \; \forall y, \; |\widehat{F}_Y^\chi(y) - \widehat{F}_Y^\chi(t_\alpha(\chi))| \le \delta(\epsilon) \Rightarrow |y - t_\alpha(\chi)| \le \epsilon.$$

By taking $y = \widehat{t}_\alpha(\chi)$, one arrives finally at the following result

$$\exists \delta(\epsilon) > 0, \; P(|\widehat{t}_\alpha(\chi) - t_\alpha(\chi)| > \epsilon) \le P(|\widehat{F}_Y^\chi(\widehat{t}_\alpha(\chi)) - \widehat{F}_Y^\chi(t_\alpha(\chi))| > \delta(\epsilon))$$
$$= P(|F_Y^\chi(t_\alpha(\chi)) - \widehat{F}_Y^\chi(t_\alpha(\chi))| > \delta(\epsilon)),$$

which, combined with the pointwise almost complete convergence of the kernel c.d.f. estimate (see Lemma 6.5), leads to (6.52). □

6.2.5 Complements on Conditional Distribution Estimation

The results stated before, as well for quantiles as for mode estimation, are based on some previous auxiliary results about the estimation of the conditional probability distribution of Y given $\mathcal{X}$. Because both functional conditional c.d.f. and functional conditional density nonlinear operators can also be interesting by themselves (and not only because of their possible applications for mode or quantile settings), this section is devoted to a general presentation of several results in these settings. Precisely, complete convergence results for kernel estimates of the functional conditional c.d.f. F_Y^χ (see Proposition 6.9) and of the conditional density f_Y^χ (see Proposition 6.10) are stated. Some of the results to be presented below were already stated while some other ones can be proved by following quite closed arguments. For these reasons, some steps of the proofs will be given in a short fashion.

The first proposition is stating complete convergence, both pointwisely and uniformly over some compact set, of the kernel functional conditional c.d.f. estimate $\widehat{F}_Y^{\mathcal{X}}$ under a continuity type model for the true conditional c.d.f. $F_Y^{\mathcal{X}}$.

Proposition 6.9. *i) Under the conditions (5.17), (6.2), (6.3) and (6.14), we have for any fixed real point y:*

$$\lim_{n \to \infty} F_Y^{\mathcal{X}}(y) = \widehat{F}_Y^{\mathcal{X}}(y), \quad a.co. \tag{6.53}$$

ii) If in addition the bandwidth g satisfies for some $\zeta > 0$ the condition $\lim_{n \to \infty} g n^\zeta = \infty$, then for any compact subset $S \subset \mathbb{R}$ we have:

$$\lim_{n \to \infty} \sup_{y \in S} |F_Y^{\mathcal{X}}(y) - \widehat{F}_Y^{\mathcal{X}}(y)| = 0, \quad a.co. \tag{6.54}$$

Proof. The proof uses the same notation as for Lemma 6.5 above. Because result i) was proved in Lemma 6.5, it only remains to prove part ii). Use (6.18) again, and note that because the term $\widehat{r}_1(\chi)$ does not depend on $y \in S$ it can be treated exactly as in the proof of Lemma 6.5. Finally, the only things to be proved are both following results:

$$\lim_{n \to \infty} \sup_{y \in S} |\mathbb{E}\widehat{r}_3(\chi, y) - F_Y^{\mathcal{X}}(y)| = 0, \tag{6.55}$$

and

$$\lim_{n \to \infty} sup_{y \in S} |\widehat{r}_3(\chi, y) - \mathbb{E}\widehat{r}_3(\chi, y)| = 0, \quad a.co. \tag{6.56}$$

Because K_0 is supported on $[-1, +1]$ and because g tends to zero, the uniform continuity property of $F_Y^{\mathcal{X}}$ over the compact set S, allows us to write that:

$$\lim_{n \to \infty} \sup_{y \in S} \sup_{v \in [-1, +1]} 1_{B(\chi, h)}(\mathcal{X}) |F_Y^{\mathcal{X}}(y - vg) - F_Y^{\mathcal{X}}(y)| = 0.$$

This last result, combined with (6.21) and (6.23), is enough to prove (6.55).

It just remains now to check the result (6.56). Using the compactness of S, we can write that $S \subset \bigcup_{k=1}^{z_n} S_k$ where $S_k = (t_k - l_n, t_k + l_n)$ and where l_n and z_n can be chosen such that:

$$l_n = C z_n^{-1} \sim C n^{-\zeta}. \tag{6.57}$$

Taking $t_y = \arg\min_{t \in \{t_1, \dots, t_{z_n}\}} |y - t|$, we have

$$\frac{1}{\widehat{r}_1(\chi)} \sup_{y \in S} |\widehat{r}_3(\chi, y) - \mathbb{E}\widehat{r}_3(\chi, y)| = D_1 + D_2 + D_3, \tag{6.58}$$

where

$$D_1 = \frac{1}{\widehat{r}_1(\chi)} \sup_{y \in S} |\widehat{r}_3(\chi, y) - \widehat{r}_3(\chi, t_y)|,$$

$$D_2 = \frac{1}{\widehat{r}_1(\chi)} \sup_{y \in S} |\widehat{r}_3(\chi, t_y) - \mathbb{E}\widehat{r}_3(\chi, t_y)|,$$

$$D_3 = \frac{1}{\widehat{r}_1(\chi)} \sup_{y \in S} |\mathbb{E}\widehat{r}_3(\chi, t_y) - \mathbb{E}\widehat{r}_3(\chi, y)|.$$

Because $K_0 = H^{(1)}$ is assumed to be of type 0, the kernel H is differentiable with a bounded derivative. This implies in particular that H is Hölder continuous with order 1, and allows us to write directly:

$$|\widehat{r}_3(\chi, y) - \widehat{r}_3(\chi, t_y)| = \frac{1}{n} \sum_{i=1}^{n} \Delta_i |H(g^{-1}(y - Y_i)) - H(g^{-1}(t_y - Y_i))|$$

$$\leq \frac{C}{n} \sum_{i=1}^{n} \Delta_i \frac{|y - t_y|}{g}$$

$$\leq C\widehat{r}_1(\chi)\frac{l_n}{g}, \qquad (6.59)$$

and the condition imposed on g, together with (6.57), lead directly to:

$$\lim_{n \to \infty} D_1 = 0. \qquad (6.60)$$

With similar arguments, we can show that

$$|\mathbb{E}\widehat{r}_3(\chi, t_y) - \mathbb{E}\widehat{r}_3(\chi, y)| \leq C\frac{l_n}{g}.$$

Once more time, the successive use of Lemma 6.3-ii and Proposition A.6-i allow us to get:

$$\lim_{n \to \infty} D_3 = 0, \ a.co. \qquad (6.61)$$

Looking now at the term D_2 we can write for any $\epsilon > 0$:

$$P(\sup_{y \in S} |\widehat{r}_3(\chi, t_y) - \mathbb{E}\widehat{r}_3(\chi, t_y)| > \epsilon) = P(\max_{j=1...z_n} |\widehat{r}_3(\chi, t_j) - \mathbb{E}\widehat{r}_3(\chi, t_j)| > \epsilon)$$

$$\leq z_n \max_{j=1...z_n} P(|\widehat{r}_3(\chi, t_j) - \mathbb{E}\widehat{r}_3(\chi, t_j)| > \epsilon).$$

By using the result (6.28), one gets directly for any $y \in S$:

$$P\left(|\widehat{r}_3(\chi, t_y) - \mathbb{E}\widehat{r}_3(\chi, t_y)| > \epsilon_0 \sqrt{\frac{\log n}{n\varphi_\chi(h)}}\right) = O(z_n n^{-C\epsilon_0}),$$

and, (6.57) with ϵ_0 large enough, leads us to:

$$\sum_{n=1}^{\infty} P \left(\sup_{y \in S} |\widehat{r}_3(\chi, t_y) - \mathbb{E}\widehat{r}_3(\chi, t_y)| > \epsilon_0 \sqrt{\frac{\log n}{n \varphi_\chi(h)}} \right) < \infty.$$

The denominator can be treated by using Lemma 6.3-ii together with Proposition A.6-ii, and so the term D_2 is such that:

$$\lim_{n \to \infty} D_2 = O_{a.co.} \left(\sqrt{\frac{\log n}{n \varphi_\chi(h)}} \right).$$

Finally, the proof is finished by using (6.58), (6.60), (6.61) and (6.62), since one arrives at

$$\frac{1}{\widehat{r}_1(\chi)} \sup_{y \in S} |\widehat{r}_3(\chi, y) - \mathbb{E}\widehat{r}_3(\chi, y)| = O_{a.co.} \left(\sqrt{\frac{\log n}{n \varphi_\chi(h)}} \right). \qquad (6.62)$$

which is a stronger result than the claimed one (6.56). □

The next proposition states complete convergence, both pointwisely and uniformly over some compact set, of the kernel functional conditional density estimate $\widehat{f}_Y^\chi$ under a continuity-type model.

Proposition 6.10. *i) Under the conditions (5.21), (6.2), (6.3) and (6.14), we have for any fixed real number y:*

$$\lim_{n \to \infty} f_Y^\chi(y) = \widehat{f}_Y^\chi(y), \quad a.co. \qquad (6.63)$$

ii) If in addition (6.31) holds, then we have for any compact $S \subset \mathbb{R}$:

$$\lim_{n \to \infty} \sup_{y \in S} |f_Y^\chi(y) - \widehat{f}_Y^\chi(y)| = 0, \quad a.co. \qquad (6.64)$$

Proof. It suffices to note that $f_Y^\chi = F_Y^{\chi(1)}$ to see that the result (6.63) is a special case of Lemma 6.15 to be stated later on (apply (6.80) with $l = 1$). On the other hand, the result (6.64) was exactly the one given by Lemma 6.7. □

6.3 Rates of Convergence

6.3.1 Regression Estimation

To complete the consistency property (see Theorem 6.1) stated for the functional kernel regression estimate $\widehat{r}(\chi)$ defined in (5.23), this section states the rate of pointwise almost complete convergence. To do that, we will consider a

Lipschitz-type model for r. As we will see, this will allow us to state precisely the behaviour of the bias and then to derive the rate of convergence.

Theorem 6.11. *Under the Lipschitz-type model (5.12) with the probability condition (6.2), if the estimator verifies (6.3) and if the response variable Y satisfies (6.4), then we have:*

$$\widehat{r}(\chi) - r(\chi) \;=\; O\left(h^\beta\right) + O_{a.co.}\left(\sqrt{\frac{\log n}{n\,\varphi_\chi(h)}}\right). \qquad (6.65)$$

Proof. The demonstration of this result follows step by step the proof of Theorem 6.1, by using Lemma 6.3 together with Lemma 6.12 below.□

Lemma 6.12. *Under (5.12) and (6.3) we have:*

$$r(\chi) - \mathbb{E}\widehat{r}_2(\chi) \;=\; O\left(h^\beta\right).$$

Proof. Following the begining of the proof of Lemma 6.2, we have:

$$r(\chi) - \mathbb{E}\widehat{r}_2(\chi) \;=\; \mathbb{E}\left((r(\chi) - r(\boldsymbol{\mathcal{X}}_1))\,\Delta_1\right).$$

Using the Lipschitz's property of r, it becomes:

$$|r(\chi) - \mathbb{E}\widehat{r}_2(\chi)| \;\leq\; C\,\mathbb{E}\left(d(\chi, \boldsymbol{\mathcal{X}}_1)^\beta \Delta_1\right),$$

Now, because the support of the kernel function K is $[0,1]$ and since $\mathbb{E}\Delta_1 = 1$, we have

$$|r(\chi) - \mathbb{E}\widehat{r}_2(\chi)| \;\leq\; C\,h^\beta. \;\square$$

6.3.2 Conditional Median Estimation

In Theorem 6.4 it was been shown that the kernel estimate $\widehat{m}(\chi)$ of the conditional functional median $m(\chi)$ was consistent as long as the underlying nonparametric functional model for the conditional c.d.f. F_Y^χ was of continuity-type. The aim of this section is to show how an additional smoothness assumption on the function $\widehat{F}_Y^\chi$ will allow us to state precisely the rates of convergence. As in regression setting (see Section 6.3.1), it will be seen in Lemma 6.14 below that a Lipschitz-type model is enough to state precisely the rate of convergence of the kernel conditional c.d.f. estimate $\widehat{F}_Y^\chi$. However, the behaviour of the median estimation depends on the flatness of the conditional c.d.f. F_Y^χ around the point $m(\chi)$, and this can be controlled in a standard way by means of the number of derivatives of F_Y^χ vanishing at point $m(\chi)$. This is done by assuming that $F_Y^\chi(.)$ is j-times continuously differentiable around $m(\chi)$ with:

$$\begin{cases} F_Y^{\mathcal{X}(l)}(m(\chi)) = 0, \forall l = 1, \ldots j-1, \\ \qquad \text{and} \\ F_Y^{\mathcal{X}(j)}(m(\chi)) > 0. \end{cases} \qquad (6.66)$$

Note that throughout this section, all the derivatives are taken with respect to the real variable y. Because this approach needs to have at hand estimates of higher order derivatives of $F_Y^{\mathcal{X}}(.)$ (see Lemma 6.15 below), the following additional assumption on the unfunctional component of the estimate is necessary:

$$\begin{cases} \lim_{n \to \infty} \dfrac{\log n}{n \, g^{2j-1} \varphi_\chi(h)} = 0, \\ \qquad \text{and} \\ H \text{ is } j\text{-times continuously differentiable.} \end{cases} \qquad (6.67)$$

Theorem 6.13. *Under the Lipschitz-type model defined by (5.18) and (6.66), if the functional variable $\mathcal{X}$ satisfies (6.2), and if the kernel estimate satisfies (6.3), (6.14) and (6.67), then we have*

$$\widehat{m}(\chi) - m(\chi) = O\left(\left(h^\beta + g^\beta \right)^{\frac{1}{j}} \right) + O_{a.co.}\left(\left(\frac{\log n}{n \varphi_\chi(h)} \right)^{\frac{1}{2j}} \right). \qquad (6.68)$$

Proof. Taylor expansion of the function $\widehat{F}_Y^{\mathcal{X}}$ leads to the existence of some m^* between $\widehat{m}(\chi)$ and $m(\chi)$ such that:

$$\widehat{F}_Y^{\mathcal{X}}(m(\chi)) - \widehat{F}_Y^{\mathcal{X}}(\widehat{m}(\chi)) = \sum_{l=1}^{j-1} \frac{(m(\chi) - \widehat{m}(\chi))^l}{l!} \widehat{F}_Y^{\mathcal{X}(l)}(m(\chi))$$
$$+ \frac{(m(\chi) - \widehat{m}(\chi))^j}{j!} \widehat{F}_Y^{\mathcal{X}(j)}(m^*).$$

Because of (6.66), this can be rewritten as:

$$\widehat{F}_Y^{\mathcal{X}}(m(\chi)) - \widehat{F}_Y^{\mathcal{X}}(\widehat{m}(\chi)) = \sum_{l=1}^{j-1} \frac{(m(\chi) - \widehat{m}(\chi))^l}{l!} \left(\widehat{F}_Y^{\mathcal{X}(l)}(m(\chi)) - F_Y^{\mathcal{X}(l)}(m(\chi)) \right)$$
$$+ \frac{(m(\chi) - \widehat{m}(\chi))^j}{j!} \widehat{F}_Y^{\mathcal{X}(j)}(m^*).$$

Because $\widehat{F}_Y^{\mathcal{X}}(\widehat{m}(\chi)) = F_Y^{\mathcal{X}}(m(\chi)) = 1/2$, we have

$$\left(m(\chi) - \widehat{m}(\chi) \right)^j \widehat{F}_Y^{\mathcal{X}(j)}(m^*) = O\left(\widehat{F}_Y^{\mathcal{X}}(m(\chi)) - F_Y^{\mathcal{X}}(m(\chi)) \right)$$
$$+ O\left(\sum_{l=1}^{j-1} (m(\chi) - \widehat{m}(\chi))^l (\widehat{F}_Y^{\mathcal{X}(l)}(m(\chi)) - F_Y^{\mathcal{X}(l)}(m(\chi))) \right).$$

By combining the results of Lemma 6.15 and Theorem 6.4, together with the fact that m^* is lying between $m(\chi)$ and $\widehat{m}(\chi)$, it follows that

$$\lim_{n\to\infty} \widehat{F}_Y^{\chi(j)}(m^*) = F_Y^{\chi(j)}(m(\chi)), \ a.co.$$

Because the second part of assumption (6.66) insures that this limit is not 0, it follows by using Proposition A.6-ii that:

$$\left(m(\chi) - \widehat{m}(\chi)\right)^j = O_{a.co.}\left(\widehat{F}_Y^\chi(m(\chi)) - F_Y^\chi(m(\chi))\right)$$
$$+ O_{a.co.}\left(\sum_{l=1}^{j-1}(m(\chi) - \widehat{m}(\chi))^l(\widehat{F}_Y^{\chi(l)}(m(\chi)) - F_Y^{\chi(l)}(m(\chi)))\right).$$

Because of (6.66), for all l in $\{0, 1, \ldots, j\}$ and for all y in a neighborhood of $m(\chi)$, it exists m^* between y and m such that:

$$F_Y^{\chi(l)}(y) - F_Y^{\chi(l)}(m(\chi)) = \frac{(y - m(\chi))^{j-l}}{(j-l)!} F_Y^{\chi(j)}(m^*).$$

which implies that $F_Y^{\chi(l)}$ is Lipschitz continuous around $m(\chi)$ with order $j - l$. So, by using now Lemma 6.14 together with Lemma 6.15, one get:

$$\left(m(\chi) - \widehat{m}(\chi)\right)^j = O_{a.co.}\left(h^\beta + g^\beta\right) + O_{a.co.}\left(\left(\frac{\log n}{n\varphi_\chi(h)}\right)^{\frac{1}{2}}\right)$$
$$+ O_{a.co.}\left(\sum_{l=1}^{j-1} A_{n,l}\right) + O_{a.co.}\left(\sum_{l=1}^{j-1} B_{n,l}\right), \quad (6.69)$$

where

$$A_{n,l} = (m(\chi) - \widehat{m}(\chi))^l\left(\frac{\log n}{n\,g^{2l-1}\varphi_\chi(h)}\right)^{\frac{1}{2}},$$

and

$$B_{n,l} = (m(\chi) - \widehat{m}(\chi))^l g^{j-l}.$$

Now, we compare the quantities $A_{n,l}$ or $B_{n,l}$ with $\left(m(\chi) - \widehat{m}(\chi)\right)^j$.

- If we suppose that there exists $l \in \{1, \ldots, j-1\}$ such that $(m(\chi) - \widehat{m}(\chi))^j = O_{a.co.}(A_{n,l})$, we can write:

$$|m(\chi) - \widehat{m}(\chi)|^j \leq C\,|m(\chi) - \widehat{m}(\chi)|^l\left(\frac{\log n}{n\,g^{2l-1}\varphi_\chi(h)}\right)^{\frac{1}{2}},$$

$$\Rightarrow |m(\chi) - \widehat{m}(\chi)|^{j-l} \leq C\left(\frac{\log n}{n\,g^{2l-1}\varphi_\chi(h)}\right)^{\frac{1}{2}},$$

$$\Rightarrow |m(\chi) - \widehat{m}(\chi)|^j \leq C\left(\frac{\log n}{n\,g^{2l-1}\varphi_\chi(h)}\right)^{\frac{j}{2(j-l)}}.$$

Therefore, due to (6.67), we have proved that for any l, $1 \leq l < j$, the following implication holds:

$$(m(\chi) - \widehat{m}(\chi))^j = O_{a.co.} (A_{n,l})$$

$$\Downarrow \qquad\qquad (6.70)$$

$$(m(\chi) - \widehat{m}(\chi))^j = O_{a.co.} \left(\sqrt{\frac{\log n}{n\,\varphi_\chi(h)}} \right).$$

- In the same way, if we suppose that there exists $l \in \{1, \ldots, j-1\}$ such that $(m(\chi) - \widehat{m}(\chi))^j = O_{a.co.} (B_{n,l})$, we can write:

$$|m(\chi) - \widehat{m}(\chi)|^j \leq C\,|m(\chi) - \widehat{m}(\chi)|^l\, g^{j-l},$$
$$\Rightarrow |m(\chi) - \widehat{m}(\chi)|^j \leq Cg^j.$$

So, we have proved that for any l, $1 \leq l < j$, the following implication holds:

$$(m(\chi) - \widehat{m}(\chi))^j = O_{a.co.} (B_{n,l})$$

$$\Downarrow \qquad\qquad (6.71)$$

$$(m(\chi) - \widehat{m}(\chi))^j = O\left(g^\beta\right).$$

Finally, by again using (6.69) together with (6.70) and (6.71), it follows that

$$\left(m(\chi) - \widehat{m}(\chi)\right)^j = O_{a.co.}\left(h^\beta + g^\beta\right) + O_{a.co.}\left(\left(\frac{\log n}{n\varphi_\chi(h)}\right)^{\frac{1}{2}} \right). \qquad (6.72)$$

This proof is now finished, at least as soon as the following lemmas are proved.$\Box$

Lemma 6.14. *Under the conditions of Theorem 6.13, we have:*

$$F_Y^\chi(y) - \widehat{F}_Y^\chi(y) = O\left(h^\beta + g^\beta\right) + O_{a.co.}\left(\left(\frac{\log n}{n\varphi_\chi(h)}\right)^{\frac{1}{2}} \right). \qquad (6.73)$$

Proof. The structure of the proof is the same as the one of Lemma 6.5, and the same notation is used. The proof is based on the decomposition (6.18). Note first that the denominators involved in this decomposition are directly treated by using Lemma 6.3-ii together with Proposition A.6-ii. So, the claimed result will be obtained as soon as the three following ones have been checked:

$$\widehat{r}_1(\chi) - 1 = O_{a.co}\left(\left(\frac{\log n}{n\varphi_\chi(h)}\right)^{\frac{1}{2}} \right), \qquad (6.74)$$

$$\mathbb{E}\widehat{r}_3(\chi, y) - F_Y^\chi(y) = O\left(g^\beta\right) + O\left(h^\beta\right), \qquad (6.75)$$

and

$$\widehat{r}_3(\chi, y) - \mathbb{E}\widehat{r}_3(\chi, y) = O_{a.co}\left(\left(\frac{\log n}{n\varphi_\chi(h)}\right)^{\frac{1}{2}} \right). \qquad (6.76)$$

- By taking

$$\epsilon = \epsilon_0 \left(\frac{\log n}{n \varphi_\chi(h)} \right)^{\frac{1}{2}},$$

inside of (6.28), one get

$$P(\widehat{r}_3(\chi, y) - \mathbb{E}\widehat{r}_3(\chi, y) > \epsilon_0 \left(\frac{\log n}{n \varphi_\chi(h)} \right)^{\frac{1}{2}}) \leq n^{-C\epsilon_0^2}, \qquad (6.77)$$

and it follows that for ϵ_0 large enough:

$$\sum_{n=1}^{\infty} P(\widehat{r}_3(\chi, y) - \mathbb{E}\widehat{r}_3(\chi, y) > \epsilon_0 \left(\frac{\log n}{n \varphi_\chi(h)} \right)^{\frac{1}{2}}) < \infty. \qquad (6.78)$$

This is enough to obtain (6.76). It remains just to check that (6.75) is true. The Lipschitz condition (5.18) allows to write that

$$\lim_{n \to \infty} \sup_{v \in [-1, +1]} 1_{B(\chi, h)}(\boldsymbol{\mathcal{X}}) |F_Y^{\boldsymbol{\mathcal{X}}}(y - vg) - F_Y^{\chi}(y)| = O\left(g^\beta \right) + O\left(h^\beta \right),$$

which can be combined together with (6.23) to lead to:

$$\lim_{n \to \infty} 1_{B(\chi, h)}(\boldsymbol{\mathcal{X}}) |\mathbb{E}(\Gamma_1(y)|\boldsymbol{\mathcal{X}}) - F_Y^{\chi}(y)| = O\left(g^\beta \right) + O\left(h^\beta \right). \qquad (6.79)$$

This, together with (6.21) and the fact that $\mathbb{E}\Delta_1 = 1$, is enough to prove (6.75).

Because the result (6.74) has been already obtained through regression study (see Lemma 6.3-ii), the proof of Lemma 6.14 is finished. $\square$

Lemma 6.15. *Let l be an integer $l \in \{1, \ldots, j\}$. Under the conditions of Theorem 6.13, we have:*

$$\lim_{n \to \infty} \widehat{F}_Y^{\chi^{(l)}}(y) = F_Y^{\chi^{(l)}}(y), \quad a.co. \qquad (6.80)$$

If in addition the function $F_Y^{\chi^{(l)}}(.)$ is Lipschitz continuous of order β_0, that is if

$$\exists C \in (0, +\infty), \forall (y, y') \in \mathbb{R}^2, |F_Y^{\chi^{(l)}}(y) - F_Y^{\chi^{(l)}}(y')| \leq C|y - y'|^{\beta_0}, \qquad (6.81)$$

then we have

$$F_Y^{\chi^{(l)}}(y) - \widehat{F}_Y^{\chi^{(l)}}(y) = O\left(h^\beta + g^{\beta_0} \right) + O_{a.co.} \left(\left(\frac{\log n}{n g^{2l-1} \varphi_\chi(h)} \right)^{\frac{1}{2}} \right). \qquad (6.82)$$

Proof. The same notation as in the proof of Lemma 6.5 is used.

- Consider the following decomposition:

$$\widehat{F}_Y^{\mathcal{X}(l)}(y) - F_Y^{\mathcal{X}(l)}(y) = \frac{1}{\widehat{r}_1(\chi)} \times$$

$$\times \left\{ \left(\widehat{r}_3^{(l)}(\chi, y) - \mathbb{E}\widehat{r}_3^{(l)}(\chi, y) \right) - \left(F_Y^{\mathcal{X}(l)}(y) - \mathbb{E}\widehat{r}_3^{(l)}(\chi, y) \right) \right\} +$$

$$+ \frac{F_Y^{\mathcal{X}(l)}(y)}{\widehat{r}_1(\chi)} \left\{ 1 - \widehat{r}_1(\chi) \right\}. \tag{6.83}$$

By following the same arguments as those invoked in the proof of Lemma 6.5, all we have to prove is that:

$$\lim_{n \to \infty} |\widehat{r}_3^{(l)}(\chi, y) - \mathbb{E}\widehat{r}_3^{(l)}(\chi, y)| = 0, \ a.co. \tag{6.84}$$

holds, and either

$$\lim_{n \to \infty} \mathbb{E}\widehat{r}_3^{(l)}(\chi, y) = F_Y^{\mathcal{X}(l)}(y), \tag{6.85}$$

when $F_Y^{\mathcal{X}(l)}$ is continuous, or

$$\mathbb{E}\widehat{r}_3^{(l)}(\chi, y) - F_Y^{\mathcal{X}(l)}(y) = O\left(g^{\beta_0}\right) + O\left(h^\beta\right), \tag{6.86}$$

when $F_Y^{\mathcal{X}(l)}$ is Lipschitz continuous, also holds.
- The proof of (6.84) is similar to the proof of of (6.20), so it is presented in a shorter way. Indeed, similar to (6.25), it is easy to write

$$\widehat{r}_3^{(l)}(\chi, y) - \mathbb{E}\widehat{r}_3^{(l)}(\chi, y) = \frac{1}{n} \sum_{i=1}^n (S_i - \mathbb{E}S_i), \tag{6.87}$$

where S_i are i.i.d. r.r.v. having zero mean and satisfying

$$|S_i| \leq C/(g^l \varphi_\chi(h)). \tag{6.88}$$

This last inequality comes from the boundedness properties of K and of all the derivatives of H. The second moment of the variables S_i can be computed in a standard way by integrating by substitution:

$$\mathbb{E}S_i^2 = \frac{1}{g^{2l}} \mathbb{E}\left(\Delta_i^2 \mathbb{E}(\Gamma_i^{(l)^2} | \mathcal{X} = \chi) \right)$$

$$= \frac{1}{g^{2l}} \mathbb{E}\left(\Delta_i^2 \int_{\mathbb{R}} H^{(l)^2}(\frac{y-u}{g}) f_Y^{\mathcal{X}}(u) du \right)$$

$$\leq \frac{1}{g^{2l-1}} \mathbb{E}\left(\Delta_i^2 \int_{\mathbb{R}} H^{(l)^2}(v) f_Y^{\mathcal{X}}(y - gv) dv \right)$$

$$\leq C \frac{1}{g^{2l-1}} \mathbb{E}\Delta_i^2$$

$$\leq C \frac{1}{g^{2l-1} \varphi_\chi(h)}, \tag{6.89}$$

the last inequality coming directly from (6.27). Note that, along these calculous, the existence of the conditional density $f_Y^{\mathcal{X}}$ is insured because of the differentiability assumptions made on the conditional c.d.f. $F_Y^{\mathcal{X}}$. We are now in position to apply the Bernstein-type exponential inequality given by Corollary A.9-ii (see Appendix A), and we get:

$$\hat{r}_3^{(l)}(\chi, y) - \mathbb{E}\hat{r}_3^{(l)}(\chi, y) = O_{a.co.}\left(\frac{\log n}{ng^{2l-1}\varphi_\chi(h)}\right)^{\frac{1}{2}}. \tag{6.90}$$

The proof of (6.84) follows.

- The proofs of (6.85) and (6.86) are also very close to those of (6.19) and (6.75), and they are therefore presented in a shorter fashion. By doing $l-1$ successive integrations by parts, and then by integrating by substitution, we arrive at:

$$\mathbb{E}(\Gamma_1^{(l)}(y)|\mathcal{X} = \chi) = \frac{1}{g^l}\int_\mathbb{R} H^{(l)}(\frac{y-u}{g})f_Y^{\mathcal{X}}(u)du$$

$$= \frac{1}{g}\int_\mathbb{R} H^{(1)}(\frac{y-u}{g})F_Y^{\mathcal{X}(l)}(u)du$$

$$= \int_\mathbb{R} K_0(v)F_Y^{\mathcal{X}(l)}(y-gv)du.$$

Note that the compact support of the kernels allows us to write that

$$F_Y^{\mathcal{X}}(y) = F_Y^\chi(y) + O(h^\beta)$$

where the quantity $O(h^\beta)$ is uniform over y. So one has

$$\mathbb{E}\hat{r}_3^{(l)}(\chi, y) - F_Y^{\chi(l)}(y)$$

$$= \mathbb{E}\left(\Delta_1\left(\int_\mathbb{R} K_0(v)(F_Y^{\mathcal{X}(l)}(y-vg) - F_Y^{\chi(l)}(y))dv\right)\right)$$

$$= \mathbb{E}\left(\Delta_1\left(\int_\mathbb{R} K_0(v)(F_Y^{\chi(l)}(y-vg) - F_Y^{\chi(l)}(y))dv\right)\right)$$

$$+ O(h^\beta). \tag{6.91}$$

Because K and K_0 are compactly supported and because g and h tend to zero, the continuity-type model for $F_Y^{\chi(l)}$ allows one to write:

$$\lim_{n\to\infty}\sup_{v\in[-1,+1]}|F_Y^{\chi(l)}(y-vg) - F_Y^{\chi(l)}(y)| = 0, \tag{6.92}$$

and the result (6.85) follows directly by combining (6.91), (6.92) and the fact that $\mathbb{E}\Delta_1 = 1$. Similarly, the result (6.86) follows by noting that the Lipschitz-type model for $F_Y^{\chi(l)}$ allows to write that

$$\sup_{v\in[-1,+1]}|F_Y^{\mathcal{X}(l)}(y-vg) - F_Y^{\chi(l)}(y)| = O\left(g^{\beta_0}\right) + O\left(h^\beta\right). \tag{6.93}$$

The proof of Lemma 6.15 is now complete. $\square$

6.3.3 Conditional Mode Estimation

In Theorem 6.6 it was shown that the kernel estimate $\widehat{\theta}(\chi)$ of the conditional functional mode $\theta(\chi)$ was consistent as long as the underlying nonparametric functional model for the conditional density function f_Y^χ was of continuity type. The aim of this section is to show how an additional smoothness assumption on the function $\widehat{f}_Y^\chi$ will allow us to state precisely the rates of convergence. As in regression setting (see Section 6.3.1) or for conditional c.d.f. estimation (see Section 6.3.2), it will be seen below that a Lipschitz-type model is enough to state precisely the rate of convergence of the kernel estimate $\widehat{f}_Y^\chi$ of the operator f_Y^χ. However, the behaviour of the mode estimation depends on the flatness of f_Y^χ around the true mode $\theta(\chi)$, and this can be controlled in a standard way by means of the number of derivatives of f_Y^χ vanishing at point $\theta(\chi)$. This is done by assuming that $f_Y^\chi(.)$ is j-times continuously differentiable around $\theta(\chi)$ with:

$$\begin{cases} f_Y^{\chi(l)}(\theta(\chi)) = 0, \forall l = 1, \ldots j-1, \\ \qquad \text{and} \\ f_Y^{\chi(l)}(\theta(\chi)) \neq 0. \end{cases} \qquad (6.94)$$

Once again, note that throughout this Section all the derivatives will be taken with respect to the real variable y.

Theorem 6.16. *Under the Lipschitz-type model defined by (5.22) and (6.94), if the functional variable $\boldsymbol{X}$ satisfies (6.2), and if the kernel estimate satisfies (6.3), (6.14), (6.31) and (6.67), then we have:*

$$\widehat{\theta}(\chi) - \theta(\chi) = O\left(\left(h^\beta + g^\beta\right)^{\frac{1}{j}}\right) + O_{a.co.}\left(\left(\frac{\log n}{n g \varphi_\chi(h)}\right)^{\frac{1}{2j}}\right). \qquad (6.95)$$

Proof. Writing a Taylor expansion of order j of the function f_Y^χ at point $\theta(\chi)$ and using the first part of condition (6.94), leads to the existence of some θ^* between $\theta(\chi)$ and $\widehat{\theta}(\chi)$ such that:

$$f_Y^\chi(\widehat{\theta}(\chi)) = f_Y^\chi(\theta(\chi)) + \frac{1}{j!} f_Y^{\chi(j)}(\theta^*)(\theta(\chi) - \widehat{\theta}(\chi))^j.$$

This result, combined with (6.33), allows one to write that:

$$f_Y^{\chi(j)}(\theta^*)(\theta(\chi) - \widehat{\theta}(\chi))^j = O\left(\sup_{s \in (\theta(\chi) - \xi, \theta(\chi) + \xi)} |\widehat{f}_Y^\chi(s) - f_Y^\chi(s)|\right). \qquad (6.96)$$

On the other hand, the continuity of the function $f_Y^{\chi(j)}$ can be written as:

$$\forall \epsilon > 0, \exists \delta_\epsilon > 0, |\theta(\chi) - \theta^*| \leq \delta_\epsilon \Rightarrow |f_Y^{\chi(j)}(\theta(\chi)) - f_Y^{\chi(j)}(\theta^*)| \leq \epsilon,$$

leading directly to:

$$\forall \epsilon > 0, \exists \delta_\epsilon > 0, P(|f_Y^{\chi(j)}(\theta(\chi)) - f_Y^{\chi(j)}(\theta^*)| > \epsilon) \leq P(|\theta(\chi) - \theta^*| > \delta_\epsilon).$$

Because the consistency result provided by Theorem 6.6 insures that θ^* tends almost completely to $\theta(\chi)$, one gets directly that:

$$\lim_{n \to \infty} f_Y^{\chi(j)}(\theta^*) = f_Y^{\chi(j)}(\theta(\chi)) \neq 0. \tag{6.97}$$

By using (6.96), (6.97) together with Proposition A.6-ii, one arrives at:

$$|\theta(\chi) - \widehat{\theta}(\chi)|^j = O_{a.co.}\left(\sup_{y \in (\theta(\chi) - \xi, \theta(\chi) + \xi)} |\widehat{f_Y^\chi}(s) - f_Y^\chi(s)| \right), \tag{6.98}$$

which is enough, after application of the Lemma 6.17 below with $l = 1$ and $S = (\theta(\chi) - \xi, \theta(\chi) + \xi)$, to complete the proof of (6.95). $\square$

Lemma 6.17. *Under the conditions of Theorem 6.16, we have for any compact $S \subset \mathbb{R}$ and for any $l = 1, \dots, j$:*

$$sup_{y \in S} |F_Y^{\chi(l)}(y) - \widehat{F}_Y^{\chi(l)}(y)| = O\left(h^\beta + g^{\beta_0} \right) +$$

$$O_{a.co.}\left(\sqrt{\frac{\log n}{ng^{2l-1}\varphi_\chi(h)}} \right). \tag{6.99}$$

Proof. Use the same notation as in the proof of Lemma 6.15.

- Using the decomposition (6.83) and invoking the same arguments as for the proof of Lemma 6.15, it turns out that all we have to prove is that both of the following results hold:

$$\frac{1}{\widehat{r}_1(\chi)} \sup_{y \in S} |\widehat{r}_3^{(l)}(\chi, y) - \mathbb{E}\widehat{r}_3^{(l)}(\chi, y)| = O_{a.co.}\left(\left(\frac{\log n}{ng^{2l-1}\varphi_\chi(h)} \right)^{\frac{1}{2}} \right), \tag{6.100}$$

and

$$\sup_{y \in S} |\mathbb{E}\widehat{r}_3^{(l)}(\chi, y) - F_Y^{\chi(l)}(y)| = O\left(g^{\beta_0} \right) + O\left(h^\beta \right). \tag{6.101}$$

- The proof of (6.101) is quite direct. For that, note first that the the limit appearing in (6.93) is uniform over y (since S is compact), and one can write:

$$\sup_{v \in [-1, +1]} \sup_{y \in S} |F_Y^{\mathcal{X}(l)}(y - vg) - F_Y^{\chi(l)}(y)| = O\left(g^{\beta_0} \right) + O\left(h^\beta \right), \tag{6.102}$$

and the result (6.101) follows directly by combining (6.91), (6.102) and the fact that $\mathbb{E}\Delta_1 = 1$.

- It remains to check that (6.100) holds. Using the compactness of S, we can write that $S \subset \bigcup_{k=1}^{z_n} S_k$ where $S_k = (t_k - l_n, t_k + l_n)$ and where l_n and z_n can be chosen such that:

$$l_n = C z_n^{-1} \sim C n^{-(l+1)\zeta - 1/2}. \tag{6.103}$$

Taking $t_y = \arg\min_{t \in \{t_1, \ldots, t_{z_n}\}} |y - t|$, we have

$$\frac{1}{\hat{r}_1(\chi)} \sup_{y \in S} |\hat{r}_3^{(l)}(\chi, y) - \mathbb{E}\hat{r}_3^{(l)}(\chi, y)| = B_1 + B_2 + B_3, \tag{6.104}$$

where

$$B_1 = \frac{1}{\hat{r}_1(\chi)} \sup_{y \in S} \left| \hat{r}_3^{(l)}(\chi, y) - \hat{r}_3^{(l)}(\chi, t_y) \right|,$$

$$B_2 = \frac{1}{\hat{r}_1(\chi)} \sup_{y \in S} \left| \hat{r}_3^{(l)}(\chi, t_y) - \mathbb{E}\hat{r}_3^{(l)}(\chi, t_y) \right|,$$

$$B_3 = \frac{1}{\hat{r}_1(\chi)} \sup_{y \in S} \left| \mathbb{E}\hat{r}_3^{(l)}(\chi, t_y) - \mathbb{E}\hat{r}_3^{(l)}(\chi, y) \right|.$$

The Hölder continuity condition on the kernel H allows us to write directly:

$$\left| \hat{r}_3^{(l)}(\chi, y) - \hat{r}_3^{(l)}(\chi, t_y) \right|$$

$$= \frac{1}{ng^l} \sum_{i=1}^{n} \Delta_i |H^{(l)}(g^{-1}(y - Y_i)) - H^{(l)}(g^{-1}(t_y - Y_i))|$$

$$\leq \frac{C}{ng^l} \sum_{i=1}^{n} \Delta_i \frac{|y - t_y|}{g}$$

$$\leq C \hat{r}_1(\chi) l_n g^{-l-1}. \tag{6.105}$$

Together with (6.103), the second part of condition (6.31) leads directly to:

$$B_1 = O\left(\sqrt{\frac{\log n}{n \, g^{2l-1} \varphi_\chi(h)}} \right). \tag{6.106}$$

Following similar steps for B_3 and using Proposition A.6-ii in addition, it holds that

$$B_3 = O_{a.co.}\left(\sqrt{\frac{\log n}{n \, g^{2l-1} \varphi_\chi(h)}} \right). \tag{6.107}$$

Looking now at the term B_2, we can write for any $\epsilon > 0$:

$$P\left(\sup_{y \in S} |\hat{r}_3^{(l)}(\chi, t_y) - \mathbb{E}\hat{r}_3^{(l)}(\chi, t_y)| > \epsilon \right)$$

$$= P\left(\max_{j=1\ldots z_n} |\hat{r}_3^{(l)}(\chi, t_j) - \mathbb{E}\hat{r}_3^{(l)}(\chi, t_j)| > \epsilon \right)$$

$$\leq z_n \max_{j=1\ldots z_n} P\left(|\hat{r}_3^{(l)}(\chi, t_j) - \mathbb{E}\hat{r}_3^{(l)}(\chi, t_j)| > \epsilon \right).$$

By using the Bernstein exponential inequality for bounded variables (see Corollary A.9-i) together with the bounds obtained in (6.88) and (6.89), we get directly for any j:

$$P\left(\left|\widehat{r}_3^{(l)}(\chi,t_j) - \mathbb{E}\widehat{r}_3^{(l)}(\chi,t_j)\right| > \epsilon_0\sqrt{\frac{\log n}{n\,g^{2l-1}\varphi_\chi(h)}}\right)$$
$$\leq C\exp\left\{-C\frac{\epsilon_0^2\log n}{(1+g^{-1})}\right\}$$
$$= O\left(n^{-C\epsilon_0^2}\right).$$

By using (6.103), one gets for ϵ_0 large enough:

$$\sum_{n=1}^{\infty}P\left(\sup_{y\in S}\left|\widehat{r}_3^{(l)}(\chi,t_y) - \mathbb{E}\widehat{r}_3^{(l)}(\chi,t_y)\right| > \epsilon_0\sqrt{\frac{\log n}{ng^{2l-1}\varphi_\chi(h)}}\right) < \infty.$$

Because its denominator is directly treated by using Lemma 6.3-ii together with Proposition A.6-ii, the term B_2 satisfies:

$$B_2 = O_{a.co.}\left(\sqrt{\frac{\log n}{n\,g^{2l-1}\varphi_\chi(h)}}\right). \tag{6.108}$$

Finally, the claimed result (6.100) follows from (6.104), (6.106), (6.107) and (6.108).

This completes the proof of this lemma. □

6.3.4 Conditional Quantile Estimation

This section is devoted to the generalization of the results given in Section 6.3.2 to the estimation of conditional functional quantiles. In other words, this section will state precisely the rate of convergence appearing in the results of Section 6.2.4. Recall that the conditional quantile of order α, denoted by $t_\alpha(\chi)$, and its kernel estimate $\widehat{t}_\alpha(\chi)$ are respectively defined by (5.9) and (5.28). As pointed out in Section 6.2.2, under the condition (6.14), the kernel conditional quantile estimate can be defined to be the unique solution of the equation

$$\widehat{t}_\alpha(\chi) = \widehat{F}_Y^{\chi-1}(\alpha). \tag{6.109}$$

Analogously with median estimation, the following assumption is needed to control the flatness of the conditional c.d.f. around the quantile to be estimated.

$$\begin{cases} F_Y^{\chi(l)}(t_\alpha(\chi)) = 0, \forall l = 1,\ldots j-1, \\ \qquad\text{and} \\ F_Y^{\chi(j)}(t_\alpha(\chi)) > 0. \end{cases} \tag{6.110}$$

Theorem 6.18. *Let $\alpha \in (0,1)$. Under the Lipschitz-type model defined by (5.17) and (6.110), if the functional variable $\mathcal{X}$ satisfies (6.2), and if the kernel estimate satisfies (6.3), (6.14), (6.67) and (6.110), then we have*

$$\widehat{t}_\alpha(\chi) - t_\alpha(\chi) = O\left(\left(h^\beta + g^\beta\right)^{\frac{1}{j}}\right) + O_{a.co.}\left(\left(\frac{\log n}{n\varphi_\chi(h)}\right)^{\frac{1}{2j}}\right). \quad (6.111)$$

Proof. The proof is similar to the proof of Theorem 6.13 , and therefore it will be presented more briefly. Taylor expansion of the function $\widehat{F}_Y^\chi$ leads to the existence of some t^* between $\widehat{t}_\alpha(\chi)$ and $t_\alpha(\chi)$ such that:

$$\left(t_\alpha(\chi) - \widehat{t}_\alpha(\chi)\right)^j \widehat{F}_Y^{\chi(j)}(t^*) = O\left(\widehat{F}_Y^\chi(t_\alpha(\chi)) - F_Y^\chi(t_\alpha(\chi))\right)$$

$$+ O\left(\sum_{l=1}^{j-1}(t_\alpha(\chi) - \widehat{t}_\alpha(\chi))^l(\widehat{F}_Y^{\chi(l)}(t_\alpha(\chi)) - F_Y^{\chi(l)}(t_\alpha(\chi)))\right).$$

By combining the results of Lemma 6.15 and Theorem 6.4, together with the fact that t^* is between $t_\alpha(\chi)$ and $\widehat{t}_\alpha(\chi)$, it follows that

$$\lim_{n\to\infty} \widehat{F}_Y^{\chi(j)}(t^*) = F_Y^{\chi(j)}(t_\alpha(\chi)), \; a.co.$$

Because the second part of assumption (6.110) insures that this limit is not 0, it follows by using Proposition A.6-ii that:

$$\left(t_\alpha(\chi) - \widehat{t}_\alpha(\chi)\right)^j = O_{a.co.}\left(\widehat{F}_Y^\chi(t_\alpha(\chi)) - F_Y^\chi(t_\alpha(\chi))\right)$$

$$+ O_{a.co.}\left(\sum_{l=1}^{j-1}(t_\alpha(\chi) - \widehat{t}_\alpha(\chi))^l(\widehat{F}_Y^{\chi(l)}(t_\alpha(\chi)) - F_Y^{\chi(l)}(t_\alpha(\chi)))\right).$$

Using now the results of Lemma 6.14 and Lemma 6.15, one get:

$$\left(t_\alpha(\chi) - \widehat{t}_\alpha(\chi)\right)^j = O_{a.co.}\left(h^\beta + g^\beta\right) + O_{a.co.}\left(\left(\frac{\log n}{n\varphi_\chi(h)}\right)^{\frac{1}{2}}\right)$$

$$+ O_{a.co.}\left(\sum_{l=1}^{j-1}A_{n,l}^*\right) + O_{a.co.}\left(\sum_{l=1}^{j-1}B_{n,l}^*\right), \quad (6.112)$$

where

$$A_{n,l}^* = (t_\alpha(\chi) - \widehat{t}_\alpha(\chi))^l\left(\frac{\log n}{n\, g^{2l-1}\varphi_\chi(h)}\right)^{\frac{1}{2}},$$

and

$$B_{n,l}^* = (t_\alpha(\chi) - \widehat{t}_\alpha(\chi))^l g^{j-l}.$$

This, combined with the results (6.70) and (6.71) by replacing $m(\chi)$ (resp. $\widehat{m}(\chi)$) with $t_\alpha(\chi)$ (resp. $\widehat{t}_\alpha(\chi)$), allows us to complete this proof. □

6.3.5 Complements on Conditional Distribution Estimation

This short section is devoted to a general presentation of several results concerning the rate of convergence for kernel estimates of the functional conditional c.d.f. F_Y^χ (see Proposition 6.19) and of the conditional density f_Y^χ, when it exists (see Proposition 6.20). Most (but not all) of these results have already been proved when dealing with conditional mode or quantile estimation.

Proposition 6.19. *i) Under the conditions (5.18), (6.66), (6.2), (6.3) and (6.14), we have for any fixed real number y:*

$$F_Y^\chi(y) - \widehat{F}_Y^\chi(y) = O\left(h^\beta + g^\beta\right) + O_{a.co.}\left(\left(\frac{\log n}{n\varphi_\chi(h)}\right)^{\frac{1}{2}}\right).$$

ii) If in any adddition the bandwidth g satisfies for some $a > 0$ the condition $\lim_{n\to\infty} gn^a = \infty$, then for any compact subset $S \subset \mathbb{R}$ we have:

$$\sup_{y\in S}|F_Y^\chi(y) - \widehat{F}_Y^\chi(y)| = O\left(h^\beta + g^\beta\right) + O_{a.co.}\left(\left(\frac{\log n}{n\varphi_\chi(h)}\right)^{\frac{1}{2}}\right).$$

Proof. The result i) was already stated by Lemma 6.14. It remains just to prove ii). For that use (6.18) again. Because the term $\widehat{r}_1(\chi)$ does not depend on $y \in S$ it can be treated exactly as in the proof of Lemma 6.5. Finally, the only things to be proved are the following results:

$$\sup_{y\in S}|\mathbb{E}\widehat{r}_3(\chi, y) - F_Y^\chi(y)| = O\left(h^\beta + g^\beta\right), \tag{6.113}$$

and

$$\frac{1}{\widehat{r}_1(\chi)}\sup_{y\in S}|\widehat{r}_3(\chi, y) - \mathbb{E}\widehat{r}_3(\chi, y)| = O_{a.co.}\left(\left(\frac{\log n}{n\varphi_\chi(h)}\right)^{\frac{1}{2}}\right).$$

Indeed, the last result was already stated in (6.62), and it just remains to show that (6.113) holds. Because K_0 is supported on $[-1, +1]$, the Lipschitz continuity property of F_Y^χ allows us to write that:

$$\sup_{y\in S}\sup_{v\in[-1,+1]} 1_{B(\chi,h)}(\boldsymbol{\mathcal{X}})|F_Y^{\boldsymbol{\mathcal{X}}}(y - vg) - F_Y^\chi(y)| = O\left(h^\beta + g^\beta\right).$$

This last result, combined with (6.21) and (6.23), is enough to prove (6.113).□

Proposition 6.20. *i) Under the conditions (5.22), (6.2), (6.3) and (6.14), we have for any fixed real number y:*

$$f_Y^X(y) - \widehat{f}_Y^X(y) = O\left(h^\beta + g^\beta\right) + O_{a.co.}\left(\left(\frac{\log n}{ng\varphi_X(h)}\right)^{\frac{1}{2}}\right).$$

ii) If in addition (6.31) holds, then we have for any compact subset $S \subset \mathbb{R}$:

$$\sup_{y \in S}|f_Y^X(y) - \widehat{f}_Y^X(y)| = O\left(h^\beta + g^\beta\right) + O_{a.co.}\left(\left(\frac{\log n}{ng\varphi_X(h)}\right)^{\frac{1}{2}}\right).$$

Proof. This proposition is just a special case (with $l = 1$ and $\beta = \beta_0$) of results that have already been proved along the previous calculations (see the second part of Lemma 6.15 and Lemma 6.17). □

6.4 Discussion, Bibliography and Open Problems

6.4.1 Bibliography on Nonparametric Functional Prediction

The aim of this short subsection is to place the results presented above in their bibliographic context. Moreover, because nonparametric statistics for functional variables is a quite recent field of statistics, the existing literature is not really large and we make an up-to-date bibliographic survey of theoretical knowledge about prediction problems from infinite-dimensional independent sample variables. A complementary survey, but for dependent functional samples, will be discussed in Part IV.

The literature concerning regression from functional variables started with the paper by [FV00], in which previous formulations of the asymptotic results presented before were proposed under less general assumptions than here. This result has been extended in several directions (see [FV02] and [FV04]). Indeed, Theorem 6.1 and Theorem 6.11 can be found in [FV04] (under slightly weaker conditions that were not introduced here in order to avoid masking the main purpose of this book). Similarly, concerning the other functional estimation problems (namely, estimating conditional c.d.f., conditional density, conditional quantiles and conditional mode), the results presented earlier in this chapter were given in [FLV05] under slightly more general forms.

There is not much additional theoretical work in these fields, and as far as we know the complementary bibliography only concerns regression setting. At this point, let us mention the recent contributions by [FMV04] and [RV05b] investigating asymptotic expansions for quadratic errors of the kernel functional regression estimate, the paper by [FPV03] in which a similar regression model/estimate based on a single functional index approach has been proposed and the one by [AV04] in which an additional partial linear component is added to the functional nonparametric regression model.

To conclude this discussion, note that the functional nonparametric regression method can be easily extended to the case of a multivariate response just by working component by component. This opens the way for new kind of application as, for instance, land use prediction problems as described in [CFG03]. Finally, note the paper by [D02] which is the only one (at least as far as we know) to deal with nonparametric modelling in the case when both the explanatory and the response variables are functional.

6.4.2 Going Back to Finite Dimensional Setting

In the setting of finite dimensional explanatory variable, the nonparametric prediction problems investigated in this chapter have been extensively studied by many authors during the last few decades. From one side the results presented in this chapter can be seen as functional adaptations of some part of this bibliography. However, from another side, the methodology developed in this book may directly be applied for finite dimensional purposes even if the main goal is not this one. (See Chapter 13 for details). Therefore, the results presented in this chapter can be also be used as new versions of some parts of the existing finite dimensional literature. What should be absolutely noted is that all throughout this chapter there is no need to assume the existence of the density function of the explanatory variable, in such a way that a direct application of this functional methodology allows us to obtain results in unfunctional setting under weaker assumptions on the distribution of the explanatory variable than those usually introduced in the classical literature.

Concerning regression estimation in the finite dimensional setting, the literature is absolutely huge and is outside the scope of this book. However the reader could look at the synthetic presentation provided by [SV00] and to the references therein to have some finite dimensional versions of Theorem 6.1 and Theorem 6.11. One could see that, apart from the work [C84], the bibliography concerning finite dimensional kernel regression estimation always assumed a density function for the covariable.

Concerning conditional c.d.f. estimation and its application for conditional quantiles, the literature in finite dimensional setting is also quite important (even if less so than in regression setting). Key references concerning the study of the unfunctional version of the kernel conditional c.d.f. estimate are [Ro69], [S89], [C97], [BGM01], [ABH03] and [MM03] while nice synthetic presentations can be found in [KB78], [PT98] or [G02]. The more general literature concerning the kernel estimation of conditional probabilities (see for instance [C80]) has also obvious applications in conditional c.d.f. estimation. The functional results presented above again have unfunctional applications (see Chapter 13 for details) and, from this point of view, complete the classical finite dimensional literature.

There is also a wide variety of literature about estimation of conditional density and conditional mode when the explanatory variable is of finite dimension. Key references concerning the study of the un-functional versions of the kernel conditional density and mode estimates can be found in [ZL85], [SM90], [YSV94], [Y96] [BH01], [GZ03] and [deGZ03] (an additional bibliography in the setting of dependent samples will be discussed later on Chapter 11). Note that the functional results presented above again have direct unfunctional applications (see again concluding chapter for details). From this point of view, the results presented earlier in this chapter complete this classical finite dimensional literature (with the main interest being not to have to assume any density for the explanatory variable).

6.4.3 Some Tracks for the Future

One thing to keep in mind at the end of this chapter is that several classical nonparametric methods, which are quite well-known in finite dimensional settings, can be adapted with nice asymptotic properties to the functional setting. The aim of this last subsection is to highlight interesting open problems which appear from the results presented before and from the related existing bibliography. Of course there are so many open questions in these fields that we have to make some arbitary selection, and we have chosen to focus on problems for which we have in mind at least some (even if small) idea on how to attack them.

On smoothing parameter selection. Naturally, as is always the case in nonparametric estimation, the role of the smoothing parameter(s) (i.e., the bandwidth(s)), becomes prominent. From a theoretical point of view, this can be easily seen from the rates of convergence of the estimators. Looking at any of the asymptotic expansions stated in Section 6.3, one see that the rates of convergence are divided into two parts: a bias component which is increasing with the bandwidths, and a dispersion component which is decreasing as the bandwidths are growing. So, there is a real need to use bandwidths that are able to balance this trade-off between bias and dispersion, or said differently, that balance the trade-off between over, and under, smoothing of the functional operator to be estimated. Answering the following question(s) will certainly be a real challenge for the future.

Open question 1 : Bandwidth choice. *How can automatic bandwidth selection procedures be developed in regression? In conditional cdf? In conditional density?*

We support the idea that most of the knowledge available in finite dimensional setting could be, after suitable adaptation, of interest in functional statistics. As far as we know, this question has only been attacked in regression by [RV05] and [RV05b] in which the usual cross-validation bandwidth procedure studied by [HM85] has been transplanted to functional variables, and by [FMV04] in which the Wild Bootstrap ideas (see [M93], [HM91] and [M00]) are discussed for functional purpose. Of course, even in regression, many other techniques existing for finite dimensional variables (see [V93] for survey) could be considered for functional purposes. Even if nothing seems to have been developed for other functional predictions settings than regression, it seems to us that possible functional extensions of the existing finite dimensional literature could be thought. For conditional density and mode, it could be possible, for instance, to adapt the techniques/results given by [YSV94] or [BH01] to functional settings.

On other modes of consistency. Most of the results available in nonparametric functional prediction are stated in terms of complete convergence. This has the main interest of making them valid as well almost surely (see Appendix A) as in probability (see also concluding chapter for more details on this mode of convergence). However, as it appears clearly from any among the results presented earlier in this chapter, one drawback of this mode of convergence is to state only upper bounds for the rates of convergence without being able to specify the constants involved in the asymptotic expansions. This is not linked with the functional feature of the problem (since the same thing is well-known in finite dimensional nonparametric statistics) but is directly related to such modes of convergence. In finite dimensional frameworks, this problem is usually attacked by considering some quadratic loss function (for instance, Mean Squared Error). The specification of the constants being very helpful for many purposes, the following question is a real challenge for the future.

Open question 2: Quadratic loss. *How can Mean Squared Errors (or other quaratic loss) expansions be obtained? In functional regression? In functional conditional cdf? In functional conditional density?*

As far as we know, the only work in this direction was provided by [FMV04] in regression setting and the interesting potential of such kinds of results have been pointed out by the authors, for bandwidth selection as well as for confidence bands construction. To complete the discussion, let us mention some recent works which provide asymptotic normality for functional conditional mode ([EO05]) and functional conditional quantile ([EO05b]).

Links with other functional models. At least in the regression setting, there is a quite important literature on parametric (mainly linear) modelling for functional regression. These functional linear regression models have been popularized by [RS97] and recent developments have been proposed in [FL98], [CFS99], [FZ00], [RS02], [CFF02], [CFS03], [CFMS03], [CGS04], [CFF04], and [JS05] while recent synthetic presentations and wider bibliographical discussions can be found in [MR03], [F03], [C04] and [RS05]. One important question that was never attacked until now is the one of testing the linearity assumption for a functional model. One way to formulate this question could be the following one:

> **Open question 3 : Testing the functional linear regression model.**
> *How can the nonparametric functional regression estimate be used to check the validity of a linear (or more generally of a parametric) functional regression model?*

Based on the extensive bibliography existing in finite dimensional settings it turns out that the nonparametric estimators can be helpful for testing parametric shapes. So, we guess that such an idea could be used in further developments on functional context to propose some answer to this question. Precisely, one could reasonably expect to be able to extend to functional setting the ideas proposed by [HM93] and which have been widely picked up again for finite dimensional problems. Of course this point is also of great interest for other prediction methods than regression, but unfortunately parametric/linear functional methodologies had only received attention for regression and nothing exists on other problems (apart from the precursor work on conditional functional linear quantiles by [CCS04]).

Still keeping in mind what exists in finite dimensional setting, there is an abundant literature studying statistical models as being intermediary between parametric and purely nonparametric models. The main purpose of these models is for dimension reduction. The literature is too wide to make an exhaustive presentation here, and we just mention the key previous reference by [S85] and the general recent monograph by [HMSW04]. Our hope is that some of these ideas developed for finite dimensional settings could be helpful in the near future for answering the following question:

> **Open question 4 : Other functional regression models.** *How can the dimensionality reduction models be adapted to infinite dimensional setting?*

As far as we know, there are just two theoretical advances in this direction: a functional version of the single index model presented in [FPV03] and a functional version of the partial linear model presented in [AV04] (see also [ACEV04] for an applied motivation of a additive functional model).

Alternative to kernel techniques. In the finite dimensional setting, there exist many alternative nonparametric smoothers that could be proposed in place of the kernel smoothers. This is true for any prediction problem, including regression, conditional c.d.f., conditional density, conditional mode, and conditional quantiles, etc. These alternative approaches involve Splines, local polynomial smoothing, wavelets, and δ-sequence methods. It is impossible, and out of scope, to discuss here the bibliography on these alternative approaches but it seems to us that, once again thinking about how extend these ideas to infinite dimensional setting, local polynomial smoothing should receive special attention.

Open question 5 : On functional local polynomial alternatives to kernel. *How can the local polynomial ideas be adapted to infinite dimensional settings?*

As far as we know, there is no advance in this direction. However we have the feeling that as in regression (see [FG96] and [FG00] for general presentations), as in conditional density (see [FYT96]), as in conditional cdf and quantiles (see [M00] or [DM01]), some ideas could be extended to the functional context. The same way, the Splines smoothing techniques (see, for instance, [W90]) and more generally the reproducing kernel Hilbert spaces ideas (see [BT04]) have been recently used with functional data (see [P06]) and one can expect many further developments in this direction.

Links with small ball probabilities The connection between the asymptotic results and the concept of small ball probabilites is obvious. Indeed, the quantity $P(\mathcal{X} \in B(\chi, h))$ appears systematically in the rates of convergence through the function $\phi_\chi(h)$, and these are small ball probabilities since h tends to zero when the sample size n increases. This is obviously strongly linked with the choice of the semi-metric d, in such a way that from a statistical point of view the question should be stated as follows:

Open question 6 : Semi-metric choice. *How can we choose the semi-metric in practice?*

In fact, as we will see later, this concerns all the problems treated in this book (and not only the prediction ones). So, these notions of small ball probabilities and semi-metric choice will deserve the general and deep discussion that the reader will find in Chapter 13. Several examples of variables $\mathcal{X}$ and of semi-metric d for which the small ball probability function $\varphi_\chi(h)$ can be calculated (or at least asymptotically evaluated when h tends to 0) will be seen there. Some earlier general guidelines for answering the semi-metric selection problem will be also presented.

7

Computational Issues

This chapter is devoted to the implementation of the functional nonparametric prediction methods based on regression, conditional quantiles and conditional mode with special attention to the regression ones. It concerns mainly users/practitioners wishing to test such functional statistical techniques on their datasets. The main goal consists in presenting several routines written in S+ or R in order to make any user familiar with such statistical methods. In particular, we build procedures with automatic choice of the smoothing parameters (bandwidths), which is especially interesting for practitioners. This chapter is written to be self-contained. However, in order to make this chapter easier to understand, we recommend the reading of the "nontheoretical" Chapters 2, 3, 4 and 5. After the description of the various routines, we propose a short case study which allows one to understand how work such procedures and how they can be easily implemented. Finally, the source codes, functional datasets, descriptions of the $R/S+$ routines and guidelines for use are given with much more detail in the companion website *http://www.lsp.ups-tlse.fr/staph/npfda*.

7.1 Computing Estimators

We focus on the various functional nonparametric prediction methods. For each of them, we present the kernel estimator and its corresponding implementations through $R/S+$ subroutine. Most of the programs propose an automatic method for selecting the smoothing parameter (i.e., for choosing the bandwidths), which makes these procedures particularly attractive for practitioners. Concerning the functional nonparametric regression method, special attention is paid because it was the first one developed from an historical point of view and because it is the most popular from a general statistical point of view. Therefore, in this regression context we propose several kernel estimators with various automatic selections of the smoothing parameter.

Note that we consider only two families of semi-metrics (computed via
`semimetric.pca` or `semimetric.deriv`) described in Chapter 3, two basic
kernel functions (see routines `triangle` or `quadratic` in the companion website[1]) described in Chapter 4 and the corresponding integrated ones (see routines `integrated.quadratic` and `integrated.triangle` in the website[1]) described in Chapter 5. However, the following package of subroutines can be
viewed as a basic library; any user, according to his statistical and program-
mer's level, can increase this library by adding his own kernels, integrated
kernels or semi-metrics routines.

We recall that we focus on the prediction problem which corresponds to the
situation when we observe n pairs $(\boldsymbol{x}_i, y_i)_{i=1,...,n}$ independently and identically
distributed: $\boldsymbol{x}_i = \{\chi_i(t_1), \ldots, \chi_i(t_J)\}$ is the discretized version of the curve
$\chi_i = \{\chi_i(t); \ t \in T\}$ measured at J points $t_1, \ldots, t_J$ whereas the y_i's are
scalar responses. In addition, $\boldsymbol{d}_q(\boldsymbol{x}_i, \boldsymbol{x}_{i'})$ denotes any semi-metric (index of
proximity) between the observed curves $\boldsymbol{x}_i$ and $\boldsymbol{x}_{i'}$. So, the statistical problem
consists in predicting the responses from the curves.

7.1.1 Prediction via Regression

First, we consider the kernel estimators defined previously in (5.23). It achieves
the prediction at an observed curve $\boldsymbol{x}_{i'}$ by building a weighted average of
the y_i's for which the corresponding $\boldsymbol{x}_i$ is such that the quantity $\boldsymbol{d}_d(\boldsymbol{x}_i, \boldsymbol{x}_{i'})$
is smaller than a positive real parameter h called bandwidth. In a second
attempt, we will consider a slightly modified version in which we replace the
bandwidth h by the number k of $\boldsymbol{x}_i$'s that are taken into account to compute
the weighted average; such methods use the terminology k-Nearest Neighbours
. For both kernel and k-NN estimators, we propose various procedures, the
most basic one being the case when the user fixes himself the smoothing
parameter h or k. Any other routine achieves an automatic selection of the
smoothing parameter. So, if the practitioner wishes to test several different
bandwidths, the basic routines can be used, or, in the opposite case, let the
other routines automatically choose them.

- **Functional kernel estimator without bandwidth selection**

The main goal is to compute the quantity:

$$R^{kernel}(\boldsymbol{x}) = \frac{\sum_{i=1}^{n} y_i K\left(\boldsymbol{d}_q(\boldsymbol{x}_i, \boldsymbol{x})/h\right)}{\sum_{i=1}^{n} K\left(\boldsymbol{d}_q(\boldsymbol{x}_i, \boldsymbol{x})/h\right)},$$

where $(\boldsymbol{x}_i, y_i)_{i=1,...,n}$ are the observed pairs and $\boldsymbol{x}$ is an observed curve
at which the regression is estimated. The user has to fix the bandwidth
h, the semi-metric $\boldsymbol{d}_q(.,.)$ and the kernel function $K(.)$. The routine
`funopare.kernel` computes the quantities

---

[1] http://www.lsp.ups-tlse.fr/staph/npfda

$$R^{kernel}(z_1), \ R^{kernel}(z_2), \ldots, R^{kernel}(z_{n'}),$$

where $z_1, \ldots, z_{n'}$ is either a new set of discretized curves or the original one $(x_1, \ldots, x_n)$.

- **Functional kernel estimator with automatic bandwidth selection**

The main goal is to compute the quantity:

$$R_{CV}^{kernel}(x) \ = \ \frac{\sum_{i=1}^{n} y_i K\left(d_q(x_i, x)/h_{opt}\right)}{\sum_{i=1}^{n} K\left(d_q(x_i, x)/h_{opt}\right)},$$

where $(x_i, y_i)_{i=1,\ldots,n}$ are the observed pairs and h_{opt} is the data-driven bandwidth obtained by a cross-validation procedure:

$$h_{opt} \ = \ \arg\min_{h} CV(h)$$

where

$$CV(h) = \sum_{i=1}^{n} \left(y_i - R_{(-i)}^{kernel}(x_i)\right)^2,$$

with

$$R_{(-i)}^{kernel}(x) \ = \ \frac{\displaystyle\sum_{j=1, j\neq i}^{n} y_j K\left(d_q(x_j, x)/h\right)}{\displaystyle\sum_{j=1, j\neq i}^{n} K\left(d_q(x_j, x)/h\right)}.$$

The user has to fix the semi-metric $d_q(.,.)$ and the kernel function $K(.)$. The routine `funopare.kernel.cv` computes the quantities

$$R_{CV}^{kernel}(z_1), \ R_{CV}^{kernel}(z_2), \ldots, R_{CV}^{kernel}(z_{n'}),$$

where $z_1, \ldots, z_{n'}$ is either a new set of discretized curves or the original one $(x_1, \ldots, x_n)$.

- **Functional kernel estimator with fixed number of neighbours**

The main goal is to compute the quantity:

$$R^{kNN}(x) \ = \ \frac{\sum_{i=1}^{n} y_i K\left(d_q(x_i, x)/h_k(x)\right)}{\sum_{i=1}^{n} K\left(d_q(x_i, x)/h_k(x)\right)},$$

where $(x_i, y_i)_{i=1,\ldots,n}$ are the observed pairs and where $h_k(x)$ is a bandwidth for which there are exactly k curves among the x_i's such that $d_q(x_i, x) < h_k(x)$. The user has to fix the semi-metric $d_q(.,.)$, the kernel function $K(.)$ and the number k. The routine `funopare.knn` computes the quantities

$$R^{kNN}(z_1), \ R^{kNN}(z_2), \ldots, R^{kNN}(z_{n'}),$$

where $z_1, \ldots, z_{n'}$ is either a new set of discretized curves or the original one $(x_1, \ldots, x_n)$.

- **Kernel estimator with *global* choice of the number of neighbours**

The main goal is to compute the quantity:

$$R_{GCV}^{kNN}(\boldsymbol{x}) = \frac{\sum_{i=1}^{n} y_i K\left(d_q(\boldsymbol{x}_i, \boldsymbol{x})/h_{k_{opt}}(\boldsymbol{x})\right)}{\sum_{i=1}^{n} K\left(d_q(\boldsymbol{x}_i, \boldsymbol{x})/h_{k_{opt}}(\boldsymbol{x})\right)},$$

where $(\boldsymbol{x}_i, y_i)_{i=1,\dots,n}$ are the observed pairs and $h_{k_{opt}}(\boldsymbol{x})$ is the bandwidth corresponding to the optimal number of neighbours obtained by a cross-validation procedure:

$$k_{opt} = \arg\min_{k} GCV(k)$$

where

$$GCV(k) = \sum_{i=1}^{n} \left(y_i - R_{(-i)}^{kNN}(\boldsymbol{x}_i)\right)^2$$

with

$$R_{(-i)}^{kNN}(\boldsymbol{x}) = \frac{\sum\limits_{j=1, j\neq i}^{n} y_j K\left(d_q(\boldsymbol{x}_j, \boldsymbol{x})/h_k(\boldsymbol{x})\right)}{\sum\limits_{j=1, j\neq i}^{n} K\left(d_q(\boldsymbol{x}_j, \boldsymbol{x})/h_k(\boldsymbol{x})\right)}.$$

The term *global* selection means that we use the same number of neighbours at any curve: $h_{k_{opt}}(\boldsymbol{x})$ depends clearly on $\boldsymbol{x}$ (the bandwidth $h_{k_{opt}}(\boldsymbol{x})$ is such that only the k_{opt}-nearest neighbours of $\boldsymbol{x}$ are taken into account) but k_{opt} is the same for any curve $\boldsymbol{x}$. So, the user has to fix the semi-metric $d_q(.,.)$ and the kernel function $K(.)$. The routine funopare.knn.gcv computes the quantities

$$R_{GCV}^{kNN}(\boldsymbol{z}_1),\ R_{GCV}^{kNN}(\boldsymbol{z}_2),\dots, R_{GCV}^{kNN}(\boldsymbol{z}_{n'}),$$

where $\boldsymbol{z}_1,\dots,\boldsymbol{z}_{n'}$ is either a new set of discretized curves or the original one $(\boldsymbol{x}_1,\dots,\boldsymbol{x}_n)$.

- **Kernel estimator with *local* choice of the number of neighbours**

The main goal is to compute the quantity:

$$R_{LCV}^{kNN}(\boldsymbol{x}) = \frac{\sum_{i=1}^{n} y_i K\left(d_q(\boldsymbol{x}_i, \boldsymbol{x})/h_{k_{opt}(\boldsymbol{x}_{i_0})}\right)}{\sum_{i=1}^{n} K\left(d_q(\boldsymbol{x}_i, \boldsymbol{x})/h_{k_{opt}(\boldsymbol{x}_{i_0})}\right)},$$

where $(\boldsymbol{x}_i, y_i)_{i=1,\dots,n}$ are the observed pairs, $i_0 = \arg\min_{i=1,\dots,n} d_q(\boldsymbol{x}, \boldsymbol{x}_i)$ and $h_{k_{opt}(\boldsymbol{x}_{i_0})}$ is the bandwidth corresponding to the optimal number of neighbours at $\boldsymbol{x}_{i_0}$ obtained by:

$$k_{opt}(\boldsymbol{x}_{i_0}) = \arg\min_k \left| y_{i_0} - \frac{\sum_{i=1,i\neq i_0}^{n} y_i K\left(d_q(\boldsymbol{x}_i, \boldsymbol{x}_{i_0})/h_{k(\boldsymbol{x}_{i_0})}\right)}{\sum_{i=1,i\neq i_0}^{n} K\left(d_q(\boldsymbol{x}_i, \boldsymbol{x}_{i_0})/h_{k(\boldsymbol{x}_{i_0})}\right)} \right|.$$

The main difference from the previous estimator appears in the local aspect of the bandwidth. More precisely, the optimal number of neighbours can change from one curve to another one. This is the reason why we use the term *local* selection. The user has to fix the semi-metric $d_q(.,.)$ and the kernel function $K(.)$. The routine `funopare.knn.lcv` computes the quantities

$$R_{LCV}^{kNN}(z_1),\ R_{LCV}^{kNN}(z_2),\ldots, R_{LCV}^{kNN}(z_{n'}),$$

where $z_1,\ldots,z_{n'}$ is either a new set of discretized curves or the original one $(\boldsymbol{x}_1,\ldots,\boldsymbol{x}_n)$.

7.1.2 Prediction via Functional Conditional Quantiles

This section deals with the prediction method via the kernel estimation of the functional conditional quantile. Unlike the previous regression techniques, this kind of method introduces a smoothing parameter for the response in addition to the one needed for the curves. We have only developed the most sophisticated procedure, that is the one involving local automatic bandwidths choices. In order to reduce the computational cost, we prefer a data-driven procedure for selecting these parameters based on two learning subsamples (instead of a standard cross-validation method). Finally, the smoothing parameters are expressed in terms of k-Nearest Neighbours in a local way (i.e., the number k can differ from one unit to another).

The main goal is to compute the quantity:

$$\forall \alpha \in (0,1/2),\ t_\alpha^{kNN}(\boldsymbol{x}) = \inf_{y\in S}\left\{F_{k_{opt},\kappa_{opt}}^{kNN}(\boldsymbol{x},y) \geq 1-\alpha\right\},$$

with

$$F_{k,\kappa}^{kNN}(\boldsymbol{x},y) = \frac{\sum_{i\in I} K\left(d_q(\boldsymbol{x}_i,\boldsymbol{x})/h_k\right) H\left((y-y_i)/g_\kappa\right)}{\sum_{i\in I} K\left(d_q(\boldsymbol{x}_i,\boldsymbol{x})/h_k\right)}.$$

The $(\boldsymbol{x}_i,y_i)_{i\in I}$ are observed pairs, h_k is defined as before in regression and g_κ is the bandwidth for which there are exactly κ among the responses y_i's such that $|y_i-y|<g_\kappa$.

In order to obtain the optimal numbers of neighbours $(k_{opt}(\boldsymbol{x})$ and $\kappa_{opt}(y))$, we randomly split our learning sample into two learning subsamples:

$$(\boldsymbol{x}_{i_1},y_{i_1})_{i_1\in I_1},\ (\boldsymbol{x}_{i_2},y_{i_2})_{i_2\in I_2},\ I_1\cap I_2=\phi, I_1\cup I_2=I \text{ and } card(I_1)=[card(I)/2],$$

and we define for each $\boldsymbol{x}$, $i^* = \arg\min_{i_2 \in I_2} d_q(\boldsymbol{x}, \boldsymbol{x}_{i_2})$. Then, we compute k_{opt} and κ_{opt} as follows:

$$(k_{opt}, \kappa_{opt}) = \arg\min_{(k,\kappa)} \left| y_{i^*} - \inf_{u \in S} \left\{ F_{k,\kappa}^{kNN}(\boldsymbol{x}_{i^*}, u) \geq 1 - \alpha \right\} \right|.$$

The optimal numbers of neighbours (k_{opt} and κ_{opt}) can change from one curve to another one. This is the reason why we use the term *local* selection. So, the user has to fix the semi-metric $d_q(.,.)$, the kernel function $K(.)$ and α. The routine funopare.quantile.lcv computes the quantities

$$t_\alpha^{kNN}(\boldsymbol{z}_1),\ t_\alpha^{kNN}(\boldsymbol{z}_2), \ldots, t_\alpha^{kNN}(\boldsymbol{z}_{n'}),$$

where $\boldsymbol{z}_1, \ldots, \boldsymbol{z}_{n'}$ is either a new set of discretized curves or the original one $(\boldsymbol{x}_1, \ldots, \boldsymbol{x}_n)$.

7.1.3 Prediction via Functional Conditional Mode

We focus now on the prediction method via the kernel estimation of the functional conditional mode. As previously, the computed estimator needs the selection of two smoothing parameters in terms of k-Nearest Neighbours in a local way (the number k can differ from one unit to another one). The data-driven procedure for choosing these parameters also involves two learning subsamples; one for building the kernel estimator, one for selecting the smoothing parameters.

The main goal is to compute the quantity:

$$\theta^{kNN}(\boldsymbol{x}) = \arg\sup_{y \in S} \widehat{f}_{k_{opt}, \kappa_{opt}}^{kNN}(\boldsymbol{x}, y),$$

with

$$f_{k,\kappa}^{kNN}(\boldsymbol{x}, y) = \frac{\sum_{i \in I} K\left(d_q(\boldsymbol{x}_i, \boldsymbol{x})/h_k\right) K_0\left((y - y_i)/g_\kappa\right)}{g_\kappa \sum_{i \in I} K\left(d_q(\boldsymbol{x}_i, \boldsymbol{x})/h_k\right)}.$$

The $(\boldsymbol{x}_i, y_i)_{i \in I}$ are observed pairs, h_k and g_κ are defined as before in Section 7.1.2. Following the same steps as in Section 7.1.2, we split our sample into two learning subsamples I_1 and I_2 and we define for each $\boldsymbol{x}$, $i^* = \arg\min_{i_2 \in I_2} d_q(\boldsymbol{x}, \boldsymbol{x}_{i_2})$. Then, we compute k_{opt} and κ_{opt} as follows:

$$(k_{opt}, \kappa_{opt}) = \arg\min_{(k,\kappa)} \left| y_{i^*} - \arg\sup_{u \in S} f_{k,\kappa}^{kNN}(\boldsymbol{x}_{i^*}, u) \right|.$$

The optimal numbers of neighbours (k_{opt} and κ_{opt}) can change form one curve to another. This is the reason why we use the term *local* selection. So, the user has to fix the semi-metric $d_q(.,.)$ and the kernel function $K(.)$. The routine funopare.mode.lcv computes the quantities

$$\theta^{kNN}(\boldsymbol{z}_1),\ \theta^{kNN}(\boldsymbol{z}_2), \ldots, \theta^{kNN}(\boldsymbol{z}_{n'}),$$

where $\boldsymbol{z}_1, \ldots, \boldsymbol{z}_{n'}$ is either a new set of discretized curves or the original one $(\boldsymbol{x}_1, \ldots, \boldsymbol{x}_n)$.

7.2 Predicting Fat Content From Spectrometric Curves

7.2.1 Chemometric Data and the Aim of the Problem

This section focuses on the spectrometric curves described in Section 2.1 and partially displayed in Figure 2.1. We recall that for each unit i (among 215 pieces of finely chopped meat), we observe one spectrometric discretized curve $(\boldsymbol{x}_i)$ which corresponds to the absorbance measured at a grid of 100 wavelengths (i.e. $\boldsymbol{x}_i = (\chi_i(\lambda_1), \ldots, \chi_i(\lambda_{100}))$). Moreover, for each unit i, we have at hand its fat content y_i obtained by analytical chemical processing. The file "spectrometric.dat" contains the pairs $(\boldsymbol{x}_i, y_i)_{i=1,\ldots,215}$ and is organized as follows:

	Col 1	$\cdots$	Col j	$\cdots$	Col 100	Col 101
Row 1	$\chi_1(\lambda_1)$	$\cdots$	$\chi_1(\lambda_j)$	$\cdots$	$\chi_1(\lambda_{100})$	y_1
$\vdots$	$\vdots$	$\vdots$	$\vdots$	$\vdots$	$\vdots$	$\vdots$
Row i	$\chi_i(\lambda_1)$	$\cdots$	$\chi_i(\lambda_j)$	$\cdots$	$\chi_i(\lambda_{100})$	y_i
$\vdots$	$\vdots$	$\vdots$	$\vdots$	$\vdots$	$\vdots$	$\vdots$
Row 215	$\chi_{215}(\lambda_1)$	$\cdots$	$\chi_{215}(\lambda_j)$	$\cdots$	$\chi_{215}(\lambda_{100})$	y_{215}

The first 100 columns correspond to the 100 channel spectrum whereas the last column contains the responses. Given a new spectrometric curve $\boldsymbol{x}$, our main task is to predict the corresponding fat content $\widehat{y}$. In fact, obtaining a spectrometric curve is less expensive (in terms of time and cost) than the analytic chemistry needed for determining the percentage of fat. So, it is an important economic challenge to predict the fat content from the spectrometric curve.

   In order to highlight the performance of our functional nonparametric prediction methods, we split our original sample into two subsamples. The first one, called *learning sample*, contains the first 160 units $((\boldsymbol{x}_i, y_i)_{i=1,\ldots,160})$. The second one, called *testing sample*, contains the last 55 units $((\boldsymbol{x}_i, y_i)_{i=161,\ldots,215})$. The learning sample allows us to build the functional kernel estimators with optimal smoothing parameter(s); both the $\boldsymbol{x}_i$'s and the corresponding y_i's are used at this stage. The testing sample is useful for achieving predictions and measuring their quality; we evaluate the functional kernel estimator (obtained with the learning sample) at $\boldsymbol{x}_{161}, \ldots, \boldsymbol{x}_{215}$ ($y_{161}, \ldots, y_{215}$ being ignored) which allows us to get the predicted responses $\widehat{y}_{161}, \ldots, \widehat{y}_{215}$.

   To measure the performance of each functional prediction method, we consider

i) the distribution of the *Square Errors*: $se_i = (y_i - \widehat{y}_i)^2$, $i = 161, \ldots, 215$, and

ii) the *Empirical Mean Square Errors*: $MSE = \dfrac{1}{55} \displaystyle\sum_{i=161}^{215} se_i$.

7.2.2 Functional Prediction in Action

We have run the three routines `funopare.knn.lcv`, `funopare.mode.lcv` and `funopare.quantile.lcv` on the spectrometric dataset, corresponding to the three prediction methods: the conditional expectation (i.e. regression) method, the functional conditional mode one and the functional conditional median one. The $R/S+$ commandlines and their corresponding explanations enabling one to load the dataset, to run the subroutines and to display the results are available on the website[2]. We end this analysis by comparing these methods through the empirical Mean Square Errors (MSE) and by suggesting a substantial improvement.

The smoothness of the curves allow us to use the semi-metrics based on the derivatives. After trying some of them, it turns out that the best one is based on the second order derivatives. The results are summarized in Figure 7.1. Conditional mode and conditional median give very similar results whereas the conditional expectation seems sensitive to high values of the response. Nevertheless, the three methods give good predictions.

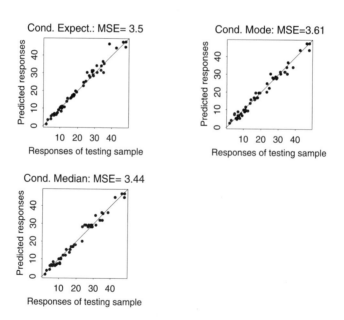

Fig. 7.1. Performance of the Three Functional Prediction Methods on Spectrometric Data

---

[2] http://www.lsp.ups-tlse.fr/staph/npfda

Because there are some differences from one method to another, one way to improve the results is to produce predictions by averaging those obtained with each method; the corresponding methodology is called *Multimethods*. As shown in Figure 7.2, the result is very interesting. There is a significant gain both in terms of mean square error and concerning the dispersion of the square error.

Fig. 7.2. Comparison Between the Three Functional Prediction Methods and the Multimethod One for Spectrometric Data

Note finally that the goal of this application was not a comparison study with other competitive methods but just to implement the three functional nonparametric prediction methods. Of course, the reader can download the spectrometric dataset on our website[3] and compare these functional statistical methods with alternative ones.

7.3 Conclusion

According to the previous results, one can say that our functional prediction methods are easy to implement and they work well for predicting fat content given spectrometric curves. In addition, subtantial improvements in terms of errors of prediction can be achieved by using the three predictive functional

---

[3] http://www.lsp.ups-tlse.fr/staph/npfda

techniques (for instance through the average of the three predictions). We will see later on that such functional methods still gives good results in the forecasting setting (i.e., time series, see Chapter 11) with another functional dataset (i.e., electricity consumption data, see Section 2.3). To conclude, let us emphasize with slight contributions, any user may easily incorporate his own semi-metrics, kernels, . . . still using the main bodies of our programs.

Part III

Nonparametric Classification of Functional Data

Classification problems motivate a large number of works in statistics. The guideline of such an area of research is to split a large collection of objects into homogeneous groups. The classification domain can be divided into two main subcategories: supervised classification and unsupervised. The supervised classification means that we have at hand a learning sample for which we know the class membership. Thus, the class structure is known a priori (i.e., observed) and the aim is to carry out a rule which allocates each object at one group. In this case, most statisticians speak about discrimination analysis whereas "supervised classification" is a terminology coming from computer science. Unsupervised classification (or cluster analysis) means that we do not observe the class membership of the considered collection of objects. This statistical problem is much more difficult because we have to define classes of objects. In such a situation, the main goal is to decide how many classes there are and how to assign each object to classes. Both supervised and unsupervised classification have been intensively studied in the multivariate case (the objects belong to a multidimensional space) and the reader can find many references in the monographs of [Go99] and [H97] whereas recent papers can be found for instance in the *Journal of Classification*. Note that functional discrimination is a prediction problem because the aim is to predict a categorical variable from a functional one. In this sense, functional discrimination could have been included in Part II. However, in the statistical literature, discrimination is generally identified with a classification problem. Therefore we voluntarily put discrimination and unsupervised classification into the same part.

Because functional data and nonparametric modelling are the guideline of our book, we propose in this part new methodologies for classifying functional data in a nonparametric way. This part is split into two chapters supervised/unsupervised classification rather than a dividing Theory/Practice as Part II. Chapter 8 focuses on the supervised classification problem. We emphasize the discriminant power of such a functional nonparametric method throughout applications to curves discrimination. Theoretical properties are easily deduced from the ones stated in Chapter 6 and so they will be presented quite briefly. Chapter 9 deals with the unsupervised classification topic. An original splitting method is developed which is based on both heuristics and theoretical advances. In addition, the splitting process is automatically stopped by a new kind of criterion and a rule classification is derived. It is worth noting that the particularity of classification makes statistical developments much more difficult in the functional framework. So, this chapter has to be seen mainly as an incitement to further investigations.

According to the spirit of this book, theoretical developments are given in self contained sections. In this way, methodological and practical aspects can be read independently of the theoretical part, which makes the reading easier both for users and for anybody else who is interested in an asymptotical outlook. Once again, one will find illustrations explaining how the methods work and how to use the corresponding $R/S+$ routines which are available on the companion website *http://www.lsp.ups-tlse.fr/staph/npfda*.

8

Functional Nonparametric
Supervised Classification

This chapter presents a nonparametric kernel method for discriminating functional data. Because theoretical advances are easily derived from those obtained in the regression setting (see Section 8.5), this chapter emphasizes applied features. The method is described in Section 8.2, the computational issues are discussed in Section 8.3 and two case studies are reported in Section 8.4. This chapter ends with some comments and bibliographical notes.

8.1 Introduction and Problematic

Supervised classification or discrimination of functional data corresponds to the situation when we observe a f.r.v. $\mathcal{X}$ and a categorical response Y which gives the class membership of each functional object. As illustration, you can refer to the speech recognition data in Section 2.2. The log-periodograms are the observations of the f.r.v. and the class membership is defined by their corresponding phonemes. The main aim in such a setting is to give reasonable answers to the following questions: given a new functional data, can we predict its class membership? Are we able to provide a consistent rule for assigning each functional object to some homogeneous group? What do we mean by a homogeneous group? How can we measure the performance of such a classification rule?

Before going on, is the classical linear discriminant analysis operational in such a setting? The answer is no because it is well known that a large number of predictors relative to the sample size and/or highly correlated predictors (which is the case when we consider functional data) lead to a degenerated within-class covariance matrix. In this functional context, the linear discrimination analysis fails. Therefore, alternative methods have been developed.

The next Section describes an alternative methodology for building a nonparametric classification rule. This is done through a proximity measure between the functional objects and a kernel estimator of the posterior probabilities derived from the one introduced in the functional nonparametric prediction context. Section 8.3 focuses on practical aspects by giving a simple way to automatically choose both the smoothing parameter introduced in the kernel estimator and the one needed for the proximity measure. To illustrate, Section 8.4 proposes applications of such a functional nonparametric method to curves discrimination; our procedure is applied to the chemometric and speech recognition data. Section 8.5 gives some theoretical properties of our kernel estimator which are easily deduced from the asymptotic behaviour of the kernel estimator in the functional nonparametric regression setting (see for more details Sections 6.2.1 and 6.3.1). The last Section is devoted to the state of the art in this area and the bibliography therein.

8.2 Method

Let $(\mathcal{X}_i, Y_i)_{i=1,\dots,n}$ be a sample of n independent pairs, identically distributed as $(\mathcal{X}, Y)$ and valued in $E \times \overline{G} = \{1, \dots, G\}$, where (E, d) is a semi-metric vector space (i.e. $\mathcal{X}$ is a f.r.v. and d a semi-metric). In practical situations, we will use the notation (χ_i, y_i) for the observation of the pair $(\mathcal{X}_i, Y_i)$, for all i varying form 1 to n. To clarify the situation, you can keep in mind the speech recognition example (Section 2.2): the $\mathcal{X}_i$'s are the log-periodograms whereas the y_i's are the corresponding classes of phoneme $(G = 5)$.

General classification rule (Bayes rule). Given a functional object χ in E, the purpose is to estimate the G posterior probabilities

$$p_g(\chi) = P(Y = g | \mathcal{X} = \chi), \quad g \in \overline{G}.$$

Once the G probabilities are estimated $(\widehat{p}_1(\chi), \dots, \widehat{p}_G(\chi))$, the classification rule consists of assigning an incoming functional observation χ to the class with highest estimated posterior probability:

$$\widehat{y}(\chi) = \arg\max_{g \in \overline{G}} \widehat{p}_g(\chi).$$

This classification rule is also called *Bayes rule*. In order to make precise our functional discriminant method, what remains is to build a suitable kernel estimator.

Kernel estimator of posterior probabilities. Before defining our kernel-type estimator of the posterior probabilities, we remark that

$$p_g(\chi) = \mathbb{E}\left(1_{[Y=g]} | \mathcal{X} = \chi\right),$$

with $1_{[Y=g]}$ equals to 1 if $Y = g$ and 0 elsewhere. In this way, it is clear that the posterior probabilities can be expressed in terms of conditional expectations.

Therefore we can use the kernel-type estimator introduced for the prediction via conditional expectation (see Section 5.4):

$$\widehat{p}_g(\chi) \;=\; \widehat{p}_{g,h}(\chi) \;=\; \frac{\sum_{i=1}^{n} 1_{[Y_i=g]}\, K\left(h^{-1}\, d(\chi, \boldsymbol{\mathcal{X}}_i)\right)}{\sum_{i=1}^{n} K\left(h^{-1}\, d(\chi, \boldsymbol{\mathcal{X}}_i)\right)}, \qquad (8.1)$$

where K is an asymmetrical kernel (see Section 4.1.2 and Definition 4.1) and h is the bandwidth (a strictly positive smoothing parameter). This really corresponds to the regression of a dichotomous variable ($1_{[Y=g]}$) on a functional one ($\boldsymbol{\mathcal{X}}$). This kernel posterior probability estimate can be rewritten as

$$\widehat{p}_{g,h}(\chi) \;=\; \sum_{\{i:\, Y_i=g\}} w_{i,h}(\chi) \;\text{ with } w_{i,h}(\chi) = \frac{K\left(h^{-1}\, d(\chi, \boldsymbol{\mathcal{X}}_i)\right)}{\sum_{i=1}^{n} K\left(h^{-1}\, d(\chi, \boldsymbol{\mathcal{X}}_i)\right)},$$

and follows the same ideas as those introduced in the context of regression. More precisely, for computing the quantity $\widehat{p}_{g,h}(\chi)$, we use only the $\boldsymbol{\mathcal{X}}_i$'s belonging to both the class g and the ball centered at χ and of radius h:

$$\widehat{p}_{g,h}(\chi) \;=\; \sum_{i\in\mathcal{I}} w_{i,h}(\chi) \text{ where } \mathcal{I} = \{i:\, Y_i = g\} \cap \{i:\, d(\chi, \boldsymbol{\mathcal{X}}_i) < h\}. \quad (8.2)$$

The closer $\boldsymbol{\mathcal{X}}_i$ is to χ the larger the quantity $K\left(h^{-1}\, d(\chi, \boldsymbol{\mathcal{X}}_i)\right)$. Hence, the closer $\boldsymbol{\mathcal{X}}_i$ is to χ the larger the weight $w_{i,h}(\chi)$. So, among the $\boldsymbol{\mathcal{X}}_i$'s lying to the gth class, the closer $\boldsymbol{\mathcal{X}}_i$ is to χ and the larger is its effect on the gth estimated posterior probability.

Before going on let us remark that, as soon as K is nonnegative, the kernel estimator has the following interesting properties

$$i)\ \ 0 \le \widehat{p}_{g,h}(\chi) < 1,$$

$$ii)\ \ \sum_{g\in\overline{G}} \widehat{p}_{g,h}(\chi) = 1,$$

which ensure that the estimated probabilities are forming a discrete distribution. Note that the first property is obvious whereas the second one comes from the fact that $\sum_{g\in\overline{G}} 1_{[Y_i=g]} = 1$.

Choosing the bandwidth. According to the shape of our kernel estimator, it is clear that we have to choose the smoothing parameter h. To do that, a usual way for an automatic choice of h consists in minimizing a loss function *Loss* as:

$$h_{Loss} \;=\; \arg\inf_{h} Loss(h),$$

where the function *Loss* can be built from $\widehat{p}_{g,h}(\chi_i)$'s and y_i's. We use the notation h_{Loss} because the automatic choice of the tuning parameter h is strongly linked with the loss function *Loss*. Even if the methodology works with any loss function, a natural choice would be the misclassification rate. That is what will do through applications in Section 8.4. We can now give the main steps of our functional discriminant procedure. Let $\mathcal{H} \subset \mathbb{R}$ be a set of reasonable values for h and K be a given asymmetrical kernel:

Learning step

    for $h \in \mathcal{H}$

        for $i = 1, 2, \ldots, n$

            for $g = 1, 2, \ldots, G$

$$\widehat{p}_{g,h}(\chi_i) \longleftarrow \frac{\sum_{\{i':\, y_{i'}=g\}} K\left(h^{-1} d(\chi_i, \chi_{i'})\right)}{\sum_{i'=1}^{n} K\left(h^{-1} d(\chi_i, \chi_{i'})\right)}$$

            enddo

        enddo

    enddo

    $h_{Loss} \longleftarrow \arg \inf_{h \in \mathcal{H}} Loss(h)$

Predicting class membership

    Let χ be a new functional object and $\widehat{y}(\chi)$ its estimated class number:

    $\widehat{y}(\chi) \longleftarrow \arg \max_{g} \{\widehat{p}_{g,h_{Loss}}(\chi)\}.$

8.3 Computational Issues

We have to choose the bandwidth h and the semi-metric $d(.,.)$ which play a major role in the behaviour of the kernel estimator defined in (8.1). However, because h is a *continuous* real parameter, from a computational point of view, it can be more efficient to replace a choice of a real parameter among an infinite number of values with an integer parameter k (among a finite subset). A simple way to do that is to consider a k Nearest Neighbours (kNN) version of our kernel estimator. This is the aim of the next section. Section 8.3.2 defines a loss function which allows a local automatic selection of the number of neighbours, and hence of the bandwidth. Section 8.3.3 gives a detailed description of the use of the discrimination routine.

    In the remaining of this section, let $(\boldsymbol{x}_i, y_i)_{i=1,\ldots,n}$ be n observed pairs identically and independently distributed; the $\boldsymbol{x}_i$'s denote the discretized functional data whereas the y_i's are the categorical responses (class numbers). In addition, as discussed in Chapter 3, we use semi-metrics denoted by $\boldsymbol{d}$ well adapted to discretized curves.

8.3.1 kNN Estimator

The kNN estimator is one way to override the problem of selecting h among an infinite subset of positive values. Indeed, the main idea of the kNN estimator is to replace the parameter h with h_k which is the bandwidth allowing us to take into account k terms in the weighted average (8.2). More precisely, if one wishes to estimate p_g at $\boldsymbol{x}$, one may use

$$\widehat{p}_{g,k}(\boldsymbol{x}) = \frac{\sum_{\{i:\ y_i=g\}}^{n} K\left(h_k^{-1} \boldsymbol{d}(\boldsymbol{x},\boldsymbol{x}_i)\right)}{\sum_{i=1}^{n} K\left(h_k^{-1} \boldsymbol{d}(\boldsymbol{x},\boldsymbol{x}_i)\right)},$$

where h_k is a bandwidth such that

$$card\{i:\ \boldsymbol{d}(\boldsymbol{x},\boldsymbol{x}_i) < h_k\} = k.$$

It is clear that we replace the minimization problem on h over a subset of $\mathbb{R}$ with a minimization on k over a finite subset $\{1, 2, \ldots, \kappa\}$:

$$k_{Loss} \longleftarrow \arg\min_{k \in \{1,\ldots,\kappa\}} Loss(k)$$
$$h_{Loss} \longleftarrow h_{k_{Loss}},$$

where the loss function $Loss$ is now built from $\widehat{p}_{g,k}(\boldsymbol{x}_i)$'s and y_i's. From now on, we consider in practice only the kNN estimator $\widehat{p}_{g,k}$ of p_g which is easy to implement.

8.3.2 Automatic Selection of the kNN Parameter

For choosing the tuning parameter k it remains to introduce a loss function $Loss$. Among the kNN estimators defined in Section 7.1 we retain the loss function allowing us to build a local version of our kNN estimator (see Section 7.1 for kNN versions of the kernel estimator in the prediction setting).

The main goal is to compute the quantity:

$$p_g^{LCV}(\boldsymbol{x}) = \frac{\sum_{\{i:\ y_i=g\}}^{n} K\left(\boldsymbol{d}(\boldsymbol{x}_i,\boldsymbol{x})/h_{LCV}(\boldsymbol{x}_{i_0})\right)}{\sum_{i=1}^{n} K\left(\boldsymbol{d}(\boldsymbol{x}_i,\boldsymbol{x})/h_{LCV}(\boldsymbol{x}_{i_0})\right)},$$

where $i_0 = \arg\min_{i=1,\ldots,n} \boldsymbol{d}(\boldsymbol{x},\boldsymbol{x}_i)$ and $h_{LCV}(\boldsymbol{x}_{i_0})$ is the bandwidth corresponding to the optimal number of neighbours at $\boldsymbol{x}_{i_0}$ obtained by the following cross-validation procedure:

$$k_{LCV}(\boldsymbol{x}_{i_0}) = \arg\min_k LCV(k, i_0),$$

where

$$LCV(k, i_0) = \sum_{g=1}^{G} \left(1_{[y_{i_0}=g]} - p_{g,k}^{(-i_0)}(\boldsymbol{x}_{i_0})\right)^2,$$

and

$$p_{g,k}^{(-i_0)}(\boldsymbol{x}_{i_0}) = \frac{\displaystyle\sum_{\{i:\ y_i=g, i\neq i_0\}}^{n} K\left(\boldsymbol{d}(\boldsymbol{x}_i, \boldsymbol{x}_{i_0})/h_{k(\boldsymbol{x}_{i_0})}\right)}{\displaystyle\sum_{i=1, i\neq i_0}^{n} K\left(\boldsymbol{d}(\boldsymbol{x}_i, \boldsymbol{x}_{i_0})/h_{k(\boldsymbol{x}_{i_0})}\right)}.$$

The main feature of such an estimator concerns the local behaviour of the bandwidth. More precisely, the optimal number of neighbours depends on the functional point at which the kNN estimator is evaluated. This is the reason why we use the term *local* selection. Note that many other loss functions can be built as in the prediction setting (see for instance the automatic choice by GCV in Section 7.1). Now, the estimation procedure is entirely determinated as soon as a semi-metric $d(.,.)$ and a kernel function $K(.)$ are fixed.

In order to give an idea of the performance of the procedure, we include the computation of the misclassification rate for the learning sample $(\boldsymbol{x}_i, y_i)_{i=1,\ldots,n}$ (i.e. the sample of curves for which the class numbers are observed):

for $i \in \{1, 2, \ldots, n\}$
$\qquad y_i^{LCV} \longleftarrow \arg\max_{g\in\{1,\ldots,G\}} p_g^{LCV}(\boldsymbol{x}_i)$
enddo

Misclas $\longleftarrow \dfrac{1}{n} \displaystyle\sum_{i=1}^{n} 1_{[y_i \neq y_i^{LCV}]}.$

8.3.3 Implementation: R/S+ Routines

We recall that we focus on the curves discrimination problem which corresponds to the situation when we observe n pairs $(\boldsymbol{x}_i, y_i)_{i=1,\ldots,n}$ independently and identically distributed: $\boldsymbol{x}_i = \{\chi_i(t_1), \ldots, \chi_i(t_J)\}$ is the discretized version of the curve $\chi_i = \{\chi_i(t);\ t \in T\}$ measured at a grid of J points $t_1, \ldots, t_J$ whereas the y_i's are the categorical responses (class membership valued into $\{1, 2, \ldots, G\}$). So, the statistical problem consists in predicting the class membership of observed curves. The routine `funopadi.knn.lcv` computes the posterior probalities

$$p_1^{LCV}(\boldsymbol{z}_1),\ p_2^{LCV}(\boldsymbol{z}_2), \ldots, p_G^{LCV}(\boldsymbol{z}_{n'}),$$

where $\boldsymbol{z}_1, \ldots, \boldsymbol{z}_{n'}$ is either a new set of discretized curves or the original one $(\boldsymbol{x}_1, \ldots, \boldsymbol{x}_n)$. Thereafter, the procedure assigns each incoming curve to the class with highest estimated posterior probability. The details of the procedure (codes and guidelines for users) are available on the website[1].

---

[1] http://www.lsp.ups-tlse.fr/staph/npfda

8.4 Functional Nonparametric Discrimination in Action

This section proposes examples of the use of functional nonparametric discrimination which emphasize its good behaviour. To do that, we consider two functional datasets: the chemometric data (spectrometric curves) and the speech recognition data. As discussed in detail in Sections 2.1 and 2.2, these functional datasets are quite different: one contains quite smooth curves whereas the second ones are particularly rough. We will see that the nonparametric methodology works well on both of them. Both datasets and commandlines for obtaining the presented results are available with details on the companion website[1].

8.4.1 Speech Recognition Problem

We recall that we observe $n = 2000$ pairs $(\boldsymbol{x}_i, y_i)_{i=1,\dots,n}$ where the $\boldsymbol{x}_i$'s correspond to the discretized log-periodogram $(\boldsymbol{x}_i = (\chi(f_1), \chi(f_2), \dots, \chi(f_{150}))$ is the ith discretized functional data) whereas the y_i's give the class membership (five phonemes):

$$y_i \in \{1, 2, 3, 4, 5\} \quad \text{with} \quad \begin{cases} 1 \longleftrightarrow \text{``sh''} \\ 2 \longleftrightarrow \text{``iy''} \\ 3 \longleftrightarrow \text{``dcl''} \\ 4 \longleftrightarrow \text{``aa''} \\ 5 \longleftrightarrow \text{``ao''} \end{cases}$$

The dataset contains the pairs $(\boldsymbol{x}_i, y_i)_{i=1,\dots,2000}$ and is organized as follows:

	Col 1	$\cdots$	Col j	$\cdots$	Col 150	Col 151
Row 1	$\chi_1(f_1)$	$\cdots$	$\chi_1(f_j)$	$\cdots$	$\chi_1(f_{150})$	y_1
$\vdots$	$\vdots$	$\vdots$	$\vdots$	$\vdots$	$\vdots$	$\vdots$
Row i	$\chi_i(f_1)$	$\cdots$	$\chi_i(f_j)$	$\cdots$	$\chi_i(f_{150})$	y_i
$\vdots$	$\vdots$	$\vdots$	$\vdots$	$\vdots$	$\vdots$	$\vdots$
Row 2000	$\chi_{2000}(f_1)$	$\cdots$	$\chi_{2000}(f_j)$	$\cdots$	$\chi_{2000}(f_{150})$	y_{2000}

The first 150 columns correspond to the 150 frequencies whereas the last column contains the categorical responses (class number). Given a new log-periodogram $\boldsymbol{x}$, our main task is to predict the corresponding class of phoneme y^{LCV}.

To measure the performance of our functional nonparametric discrimination method, we build two samples from the original dataset. The first one, the learning sample, contains the 5×50 units $((\boldsymbol{x}_i, y_i)_{i \in \mathcal{L}}$, each group containing 50 observations). The second one is the testing sample and contains 5×50 units $((\boldsymbol{x}_{i'}, y_{i'})_{\in i' \mathcal{T}}$ with 50 observations by group). The learning sample allows us to estimate the posterior probabilities with optimal smoothing parameter (i.e.,

$p_1^{LCV}(.), \ldots, p_5^{LCV}(.)$); both the $\boldsymbol{x}_i$'s and the corresponding y_i's are used at this stage. The testing sample is useful for measuring the discriminant power of such a method; we evaluate the estimators $p_1^{LCV}(.), \ldots, p_5^{LCV}(.)$ (obtained with the learning sample) at $\{\boldsymbol{x}_{i'}\}_{i' \in \mathcal{T}}$ ($\{y_{i'}\}_{i' \in \mathcal{T}}$ being ignored) which allows us to get the predicted class membership $\{y_{i'}^{LCV}\}_{i' \in \mathcal{T}}$. It remains to compute the misclassification rate

$$Misclas_{Test} \longleftarrow \frac{1}{250} \sum_{i' \in \mathcal{T}} 1_{[y_{i'} \neq y_{i'}^{LCV}]}.$$

We repeat 50 times this procedure by building randomly 50 learning samples $\mathcal{L}_1, \ldots, \mathcal{L}_{50}$ and 50 testing samples $\mathcal{T}_1, \ldots, \mathcal{T}_{50}$. Finally, we get 50 misclassification rates $Miscla_1, \ldots, Miscals_{50}$ and the distribution of these quantities gives a good idea of the discriminant power of such a functional nonparametric supervised classification. This procedure is entirely repeated, by running the routine funopadi.knn.lcv described previously, for various semi-metrics in order to highlight the importance of such a proximity measure:

- pca-type semi-metrics (routine semimetric.pca) with a number of dimension taking its values in 4, 5, 6, 7 and 8 successively,
- pls-type semi-metrics (routine semimetric.mplsr) with a number of factors taking its values in 5, 6, 7, 8 and 9 successively,
- derivative-type semi-metrics (routine semimetric.deriv) with a number of derivatives equals to zero (classical L_2 norm).

The Figure 8.1 displays the results obtained for one splitting of the sample into learning and testing subsamples.

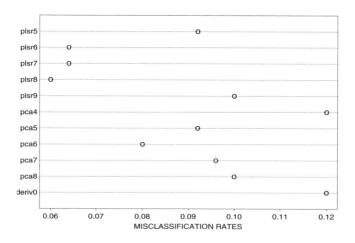

Fig. 8.1. Speech Recognition Data Discrimination: One Run

The PLS-type semi-metrics are well adapted for such a speech recognition problem.

In order to obtain more robust results, it is easy to repeat this sample splitting (for instance by using a loop) for obtaining 50 misclassification rates for each semi-metric. In this case, boxplots can be displayed as in Figure 8.2.

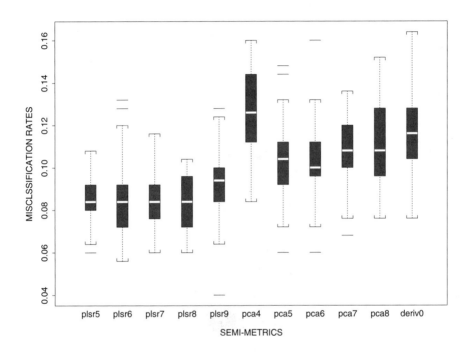

Fig. 8.2. Speech Recognition Data Discrimination: 50 runs

Note that the PLS-type semi-metrics use additional information about the categorical reponse for measuring the proximity between curves which is not the case for the PCA-type ones. Moreover, for the semi-metrics based on the derivatives, the results are given for only the L_2 metric (derivative of order zero) because they were even worse for higher order derivatives. This bad behaviour is not surprising because of the roughness of the curves. Finally, it is clear that the semi-metrics based on the multivariate PLS regression allow to obtain a good discrimination (in particular with 6 factors).

8.4.2 Chemometric Data

The used spectrometric data are described Section 7.2.1 but in a slightly modified version. Indeed, in such a discrimination setting, we have to consider a categorical response instead of a scalar one. Therefore, the observed responses $y_1, \ldots, y_{215}$ (column 101) are replaced with $y_1^*, \ldots, y_{215}^*$ where

$$\forall i = 1, \ldots, 215, \quad y_i^* = \begin{cases} 1 \text{ if } y_i < 20 \\ 2 \text{ else.} \end{cases}$$

The curves in the group labeled "1" (resp. "2") correspond to a fat content smaller (larger) than 20 %.

To measure the performance of the functional discrimination procedure, we follow the same methodology as the one used in the speech recognition problem. More precisely, we built 50 learning and testing samples (the ratio between groups being preserved) which allow us to get 50 misclassification rates. As pointed out in the regression setting (see Section 7.2.2), the smooth shape of the curves allows us to use semi-metrics based on the derivatives. We give here directly the results with the semi-metric d_2^{deriv} (see Section 3.4.3). Figure 8.3 summarizes the results. The semi-metric d_2^{deriv} leads clearly to good discrimination (the mean of the misclassification rates equals to 2 %).

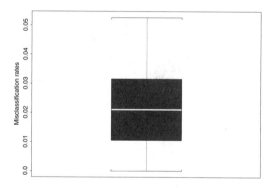

Fig. 8.3. Spectrometric Data Discrimination: 50 runs

8.5 Asymptotic Advances

As explained in introduction to Part III, the discrimination problem can be viewed as a prediction problem since it comes to estimate the conditional expectation of indicator variables (one by class). So, all asymptotic results stated in the functional context remain valid in the discrimination setting. Using the same notation as in Chapter 6, we can get the following theorems; the first one gives the pointwise almost complete convergence of the estimator of the posterior probabilities whereas the second one gives their precise rate of convergence.

Theorem 8.1. *Under the continuity-type model (i.e. $p_g \in C_E^0$) with the probability condition (6.2) and if the estimator verifies (6.3), then we have, for $g = 1, \ldots, G$:*

$$\lim_{n \to \infty} \widehat{p}_{g,h}(\chi) = p_g(\chi), \quad a.co.$$

Proof. This result is derived directly from the proof of Theorem 6.1 where the regression r is replaced with p_g and for the particular response variable $1_{[Y=g]}$. To see that, it suffices to remark that the response variable $1_{[Y=g]}$ satisfies condition (6.4). Indeed, we have

$$E\left(1_{[Y=g]}^m | \mathcal{X} = \chi\right) = P(Y = g | \mathcal{X} = \chi) \overset{def}{=} p_g(\chi).$$

Note that the continuity assumption insures that the hypothesis (6.4) is verified (where $p_g(.)$ plays the role of $\sigma_m(.)$). $\square$

Theorem 8.2. *Under the Lipschitz-type model (i.e. $p_g \in Lip_{E,\beta}$) with the probability condition (6.2) and if the estimator verifies (6.3), then we have, for $g = 1, \ldots, G$:*

$$\widehat{p}_{g,h}(\chi) - p_g(\chi) = O\left(h^\beta\right) + O_{a.co.}\left(\sqrt{\frac{\log n}{n\,\varphi_\chi(h)}}\right).$$

Proof. As before, it suffices to remark that condition (6.4) is still satisfied. So, the use of the proof of Theorem 6.11 when the regression r is replaced with the posterior probability p_g and for the particular response variable $1_{[Y=g]}$ allows us to get the result. $\square$

8.6 Additional Bibliography and Comments

As explained in the introduction to this chapter, the linear discrimination analysis fails in the functional context. Therefore, for about a decade, several statisticians investigated alternative approaches. [HBT95] proposed a regularized version of the linear discrimination analysis called *penalized discriminant analysis* whereas [ME99] developed a *generalized linear regression* approach taking into account the functional feature of the data. In order to capture nonlinear decision boundaries, [HBT94] built a *flexible discriminant analysis*. But all these methods correspond to functional parametric models according to our definitions (see definitions in Section 1.3). More recently, a nonparametric approach for curves discrimination based on a kernel density estimator of random vectors has been introduced by [HPP01], where the curves are considered as random vectors after a projection step. It is easy to see that this work corresponds to the particular case of our nonparametric functional methodology when we consider the semi-metric based on the functional principal component analysis.

Finite dimensional nonparametric discrimination was widely investigated in the last few years (see for instance [N76], [G81], [K86], [P88], [K91], [LP94], [GC04] for a non-exhaustive list of references). It is worth noting that the functional nonparametric discrimination method can be applied directly to finite dimensional context (see the general discussion in Section 13.5). Theoretical results given before can also be seen as a slight extension of similar ones already existing in the above-mentioned literature.

The method presented in this chapter comes from [FV03]. Concerning theoretical functional framework, we did not investigate the interesting problem of consistency of our Bayes classification rule based on the kernel-type estimator of the posterior probability. However, [ABC05] studied recently the consistency of a simplified version of this classifier when the kernel K is replaced with the indicator function. More details about the notion of consistency of classification rules and universal consistency can be found in [DGL96] and we hope that further works will concern such properties. There are many other open problems which are similar to those existing in the functional regression context. In particular, the bandwidth choice (see open question 1) should be developed in the near future to provide complete theoretical support for the automatic procedure presented before. From a practical point of view, Section 8.4 gave examples of the use of our functional nonparametric approach and showed the relevance of such a discrimination method. However, in order to define precisely what "relevance" means, the previous practical cases should be completed by comparative studies with several competitive methods and datasets. However, the main goal of this book consists in proposing alternative statistical methods in a functional and nonparametric way. Note that comparative studies on the datasets presented in this book, can be found in [FV03]. Note also that the proposed method has been used in [NCA04] for a polymer discrimination problem, and it gave appealing results. But, it is clear that in the near future, other alternative methods will be proposed and new works could consist in investigating deep comparisons.

Finally, for both applied and theoretical purposes, a crucial point concerns the choice of the semi-metric. Because this is the key in any nonparametric functional problem, this will be a subject of the general discussion in Chapter 13.

9

Functional Nonparametric
Unsupervised Classification

Most of the sections presented in this chapter are readable by a very large public. Methodological, practical and computational aspects take a large share whereas theoretical developments are presented in a self-contained section. In fact, the spirit of this chapter is close to the spirit of exploratory data analysis. More precisely, when an unsupervised classification is performed, the statistician or more generally the user does not know how to validate the obtained partition. Only some additional information collected after the analysis can confirm or refute the results. So, according to their experience, the statistician will try to propose more pertinent answers to the classification problem. This is exactly what we try to do here but in a new field which concerns functional data. Heuristics and theoretical aspects are developed in a complementary way which produces an original nonparametric classification method for functional data. Note that this chapter is quite different from the other ones. From a theoretical point of view, we propose a classification method which involves the mode of the distribution of a functional random variable. This leads us to deal with the density of functional random variables and new problems emerge as soon as we focus on the asymptotic behaviour of the estimator of the "functional" mode. First results allow us to point out difficulties linked with the infinite dimensional setting. From a practical point of view, it is much more simple to solve the prediction problem than the unsupervised classification one. For all these reasons, this chapter is certainly more open than the other ones and will certainly deserve future investigation.

9.1 Introduction and Problematic

Unsupervised classification is an important domain of statistics with many applications in various fields. The aim of this chapter is to propose a nonparametric way to classify a sample of functional data into homogeneous groups. The main difference with discrimination problems (see Chapter 8) is that the group structure is unknown (we do not have any observations of some categorical response), and this makes such a statistical study more delicate. The general idea is to build a descending hierarchical method which combines functional features of the data with a nonparametric approach. More precisely, the proposed methodology performs iteratively splitting into less and less heterogeneous groups. This forces us to define what means heterogeneity for a class of functional objects. To this end, we measure the closeness between some centrality features of the distribution. The great interest of the nonparametric modelling consists in estimating such characteristics without specifying the probability distribution of the functional variable. This is a required point because the distribution generating the sample of functional data is supposedly unknown (free-distribution modelling) and even if one would specify the distribution, it would be impossible to check it. Concerning the way to split the functional data, we make a feedback between practical aspects and recent theoretical advances, which allows us to introduce a partitioning based on small ball probabilities considerations.

Concerning the organization of this chapter, we voluntary insist on methodological and practical aspects as in the discrimination chapter because most of the expectations in this domain are oriented towards the applications. However, some first asymptotical advances are given in a self-contained section in order to point out open problems in relation to the infinite dimensional feature, especially when uniform consistency is required. In addition, a feedback practice/theory will be emphasized in the proposed methodology. We start with Section 9.2 which presents functional versions of the usual notions like mean, median and mode. After that, Section 9.3 proposes a simple way to measure the heterogeneity of a sample of functional data by comparing the previous indices throughout a semi-metric. Once the heterogeneity index is defined, Section 9.4 describes a general descending hierarchical method based on a notion of gain or loss when a sample of functional data is partitioned. In particular, if we have to fix some smoothing parameters (which can appear in the estimator of the mode), an automatic selection procedure based on the maximization of the entropy is proposed. Section 9.5 presents a short case study which illustrates the easiness of both implementation and use of such a nonparametric functional classification whereas its good behaviour is emphasized. Section 9.6 studies the asymptotical behavior of the kernel estimator of the mode. These theoretical developments are quite different from those detailed in Chapter 6 because a uniform-type consistency of the density estimator is necessary. Finally, this chapter ends with a bibliographical overview which places this work in the recent literature.

Before going on, we recall that the same notation are used as before. Thus, $\mathcal{X}$ denotes a generic functional variable taking its values in the infinite dimensional semi-metric space (E, d). In addition, let $\mathcal{S} = \{\mathcal{X}_1, \ldots, \mathcal{X}_n\}$ be a sample of n variables identically and independently distributed as $\mathcal{X}$, let χ be a fixed element of E and let $\chi_1, \ldots, \chi_n$ be the functional dataset associated with the functional sample $\mathcal{X}_1, \ldots, \mathcal{X}_n$.

9.2 Centrality Notions for Functional Variables

We start by describing very standard features as mean, median and mode but for a distribution of a functional variable $\mathcal{X}$. We will see that these usual notions in the multivariate case can be extended easily to the infinite dimensional context.

9.2.1 Mean

The simplest one and the most popular is the mean. Formally, the mean of $\mathcal{X}$ is defined by

$$\mathbb{E}(\mathcal{X}) = \int_{\Omega} \mathcal{X}(\omega)\, dP(\omega),$$

where $(\Omega, \mathcal{A}, P)$ is the probability space. Once the mathematical definition is given, a "universally" and always computable well-known estimator of the mean is its empirical version:

$$\mathcal{X}_{mean,\mathcal{S}} = \frac{1}{n}\sum_{i=1}^{n}\mathcal{X}_i.$$

For instance, if E is a standard real functional space, we have the usual notion of mean curve:

$$\forall t \in \mathbb{R}, \ \mathcal{X}_{mean,\mathcal{S}}(t) = \frac{1}{n}\sum_{i=1}^{n}\mathcal{X}_i(t).$$

However, one has to use it carefully according to the shape of the data. In particular, the mean can be often non-informative when the data are rough. For instance, if one is considering any log-periodogram classes of the speech recognition dataset (see Section 8.4.1) and their corresponding functional mean displayed in Figure 9.1, it seems obvious that the global mean curve (mean over the whole sample) is oversmoothing such rough data (in comparison with the mean curves obtained for each class).

Other situations when the notion of mean curve fails, are the ones which appear when there is horizontal shift or unbalanced data due to the apparatus. Horizontal shift occurs when one considers wave forms collected by the satellite Topex/Poseidon (see [DFV04]). In such a situation, the mean makes no sense and specific analyses are needed (see discussion in Section 3.6). The case of unbalanced data can be solved by performing approximations at same measurements as soon as the design for each unit is sufficiently fine enough (see again Section 3.6). In return, spectrometric data (see Section 8.4.2) seems to be well adapted for using the mean as one can see in Figure 9.2. Except for a vertical shift, the shape of the global mean is very close to those computed for each group.

Empirical means

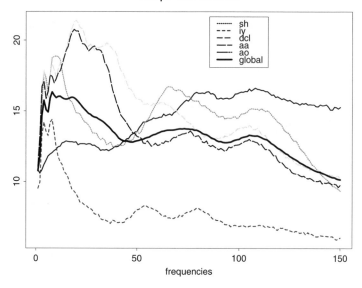

Fig. 9.1. Speech Recognition Data: Mean Curves (Global and by Group)

Empirical means

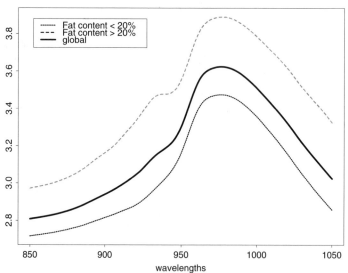

Fig. 9.2. Spectrometric Data: Mean Curves (Global and by Group)

Finally, let us note that a robust version of the mean curve (called trimmed mean) is defined in [FM01]. It is built by using a functional version of the depth notion.

9.2.2 Median

Another way to define the mean of a r.r.v X is to consider it as the solution (under existence and uniqueness) of the following minimization problem:

$$\inf_{x \in \mathbb{R}} \mathbb{E} \left(x - X \right)^2 .$$

In the same way, one can define the median of a r.r.v. X as the solution (under existence and uniqueness) of the minimization problem:

$$\inf_{x \in \mathbb{R}} \mathbb{E} \left(|x - X| \right) .$$

One of the main properties of the median is to be more robust than the mean. Indeed, it is well known that outliers can dramatically deteriorate the mean without significant consequence on the median.

It is easy to extend such ideas to the functional case (under existence) by replacing $\mathbb{R}$ with E and $|x - X|$ with $d(\chi, \boldsymbol{X})$. This leads us to introduce the following definition.

Definition 9.1. *A solution (under existence assumption) of the following minimization problem*

$$\inf_{\chi \in E} \mathbb{E} \left(d(\chi, \boldsymbol{X}) \right)$$

will be called functional median associated to the semi-metric d.

In the following, because d is fixed, we will call it simply *functional median*.

Theoretical aspects of median in Banach space have been studied by [K87] (see also [C01] for recent advances). An empirical estimator of the (functional) median is given by:

$$\boldsymbol{X}_{med} = \inf_{\chi \in E} \sum_{i=1}^{n} d(\chi, \boldsymbol{X}_i),$$

and a computable and simplest one can be obtained as follows:

$$\boldsymbol{X}_{med,\mathcal{S}} = \inf_{\chi \in \mathcal{S}} \sum_{i=1}^{n} d(\chi, \boldsymbol{X}_i),$$

where we recall that $\mathcal{S} = \{\boldsymbol{X}_1, \ldots, \boldsymbol{X}_n\}$. Contrary to $\boldsymbol{X}_{mean,\mathcal{S}}$, the semi-metric d will play a major role in computing the median. Rather than a drawback, the possibility of using various semi-metrics d can be seen as a tuning tool.

9.2.3 Mode

In finite dimensional setting, the mode is very popular in classification because it is a useful tool for depicting groups and also more robust than the mean (like the median, the mode is less sensitive to outliers than the mean). The functional extension of this notion requests that the distribution of the functional variable $\mathcal{X}$ has a density f with respect to some abstract measure μ defined on the infinite dimensional space E. This assumption is implicitly made in all the remaining. Formally, a mode θ of the probability distribution of $\mathcal{X}$ is a value that locally maximizes its density f. So, we propose the following definition:

Definition 9.2. *A solution (under existence assumption) of the following maximization problem*

$$\sup_{\chi \in \mathcal{C}} f(\chi),$$

where $\mathcal{C}$ is a subset of non-empty interior of E, will be called functional mode (implicitly associated to the semi-metric d and to the measure μ).

In order to make the mode computable, f has to be estimated and the supremum has to be taken over a finite subset of E.

Estimating the density. We consider a pseudo kernel-type functional density estimator defined by:

$$\forall \chi \in E, \ \tilde{f}(\chi) = \frac{1}{Q(K,h)} \sum_{i=1}^{n} K\left(h^{-1} d(\chi, \mathcal{X}_i)\right),$$

where $Q(K,h)$ is a positive quantity which does not depend on χ. $\tilde{f}$ is not exactly an estimator because as we will see in Section 9.6, the quantity $Q(K,h)$ is unknown. However, for mode purpose, this not a real problem since in order to maximize $\tilde{f}$ we just have to compute $\sum_{i=1}^{n} K\left(h^{-1} d(\chi, \mathcal{X}_i)\right)$, which is very easy to implement.

Estimating the mode. According to this density estimator, for estimating the mode it remains to maximize it over $\mathcal{C}$. But, from a practical point of view, it is impossible to optimize $\tilde{f}$ over so large a subset of E. Following the same ideas as in [ABC03], a simple way to override this problem consists in replacing $\mathcal{C}$ with the sample $\mathcal{S}$ and the empirical functional version θ_n of the mode θ is defined by:

$$\mathcal{X}_{mod,\mathcal{S}} = \arg\max_{\xi \in \mathcal{S}} \tilde{f}(\xi).$$

Clearly, the maximization acts over a finite set (i.e., the sample) which makes this estimation very easy to compute.

9.3 Measuring Heterogeneity

When one has at hand two features of centrality (as mean, median or mode) of a distribution of a functional variable denoted by $\mathcal{M}_{1,\mathcal{S}}$ and $\mathcal{M}_{2,\mathcal{S}}$, it is quite natural to measure the heterogeneity of the given sample $\mathcal{S}$ by comparing $\mathcal{M}_{1,\mathcal{S}}$ with $\mathcal{M}_{2,\mathcal{S}}$. More precisely, we will call Heterogeneity Index of a sample $\mathcal{S}$, denoted $HI(\mathcal{S})$, any quantity which can be expressed as follows:

$$HI(\mathcal{S}) = \frac{d(\mathcal{M}_{1,\mathcal{S}}, \mathcal{M}_{2,\mathcal{S}})}{d(\mathcal{M}_{1,\mathcal{S}}, 0) + d(\mathcal{M}_{2,\mathcal{S}}, 0)}.$$

It is important to remark that the mode will play a major role in exhibiting heterogeneity. One will always take $(\mathcal{M}_{1,\mathcal{S}}, \mathcal{M}_{2,\mathcal{S}}) = (\boldsymbol{X}_{mean,S}, \boldsymbol{X}_{mod,S})$ or eventually $(\mathcal{M}_{1,\mathcal{S}}, \mathcal{M}_{2,\mathcal{S}}) = (\boldsymbol{X}_{med,S}, \boldsymbol{X}_{mod,S})$ if the mean seems to be not very well adapted to the considered situation. Thus, we expect that the larger $HI(\mathcal{S})$ is and the more heterogeneous is the sample $\mathcal{S}$. In addition, according to the importance of the mode in this problem, first theoretical advances will be focused on it (see Section 9.6). It is obvious that this criterion plays a crucial role in the classification procedure. This is the reason why we introduce a subsampled version of $HI(\mathcal{S})$ in order to make it more robust. To do that, consider L randomly generated subsamples $\mathcal{S}^{(l)}$ ($\subset \mathcal{S}$) of same size and let the Subsampling Heterogeneity Index of $\mathcal{S}$ be defined as:

$$SHI(\mathcal{S}) = \frac{1}{L} \sum_{l=1}^{L} HI(\mathcal{S}^{(l)}).$$

Actually, the quantity $SHI(\mathcal{S})$ can be viewed as an approximation of the expectation of $HI(\mathcal{S})$.

9.4 A General Descending Hierarchical Method

This section describes a general descending hierarchical method for classifying functional data. Starting from the whole sample, if we decide to split it, we build a partition and we repeat this procedure for each performed group. So, we have to answer both important questions: is the obtained partition more pertinent/informative than the previous one? How to build the partition? One way to answer the first question consists in proposing a criterion able to compute a heterogeneity index of a partition. After that, it suffices to compare it with the heterogeneity index of the *father* sample and to deduce a splitting score in terms of gain or loss of heterogeneity. Of course, a gain large enough implies that the partition is accepted; in the opposite case, the *father* sample is a terminal leaf of the classification procedure. The next section deals with such a partitioning heterogeneity index and splitting score. Concerning the second question, as we will see in Section 9.4.2, small ball probability gives a useful tool for building a partition of a given sample.

9.4.1 How to Build a Partitioning Heterogeneity Index?

One has to decide if a current group (and the whole sample itself) deserves
to be split or not. To arrive at such a decision, we build a stop criterion
of the splitting process based on a partitioning heterogeneity index. So, let
$\mathcal{S}_1, \ldots, \mathcal{S}_K$ be a partition of $\mathcal{S}$ and define a Partitioning Heterogeneity Index
as a weighted average of the SHI of each component:

$$PHI(\mathcal{S}; \mathcal{S}_1, \ldots, \mathcal{S}_K) = \frac{1}{card(\mathcal{S})} \sum_{k=1}^{K} card(\mathcal{S}_k) \times SHI(\mathcal{S}_k).$$

Once $PHI(\mathcal{S}; \mathcal{S}_1, \ldots, \mathcal{S}_K)$ is introduced, it is easy to deduce the following
Splitting Score:

$$SC(\mathcal{S}; \mathcal{S}_1, \ldots, \mathcal{S}_K) = \frac{SHI(\mathcal{S}) - PHI(\mathcal{S}; \mathcal{S}_1, \ldots, \mathcal{S}_K)}{SHI(\mathcal{S})}.$$

If this quantity is positive, the splitting score can be expressed in terms of
a loss of heterogeneity (i.e., a global gain of homogeneity) and the splitting
is considered pertinent as soon as the splitting score is greater than a fixed
threshold. If the score is negative (i.e., a global loss of homogeneity), $\mathcal{S}$ is not
split.

9.4.2 How to Build a Partition?

A central question in the classification concerns the building of the partition.
On the other hand, according to the simple kernel estimator of the density,
we have to choose the bandwidth h. So, we propose here an original way to
perform both the partition and choose an *optimal* bandwidth.

Small ball probability and probability curves. To do that, we introduce the
quantity $P(\mathcal{X} \in B(\chi, h))$ which plays a major role in the rates of convergence in most of the asymptotic results in the functional setting (see Chapter
13 for general discussion). From a theoretical point of view, recall that h is
a positive sequence which tends to zero with the size of the sample n and
$B(\chi, h) = \{\chi' \in E, \ d(\chi, \chi') < h\}$. Therefore the terminology *small ball probability* is used for $P(\mathcal{X} \in B(\chi, h))$ (see Section 4.2 for more details). It is easy
to estimate such a small ball probability by:

$$\hat{p}_i(h) = \hat{P}(\mathcal{X} \in B(\chi_i, h)) = \frac{1}{n} \, card \{i'/d(\chi_i, \chi_{i'}) < h\}.$$

For each i, we can display $\hat{p}_i(h)$ versus h and hence we obtain n *probability
curves*. For instance, if we consider the sample of spectrometric data, we get
the left graphic of Figure 9.3 which displays the probability curves $\hat{p}_i(.)$, $i = 1, \ldots, n = 215$. The right part of this graphic shows the estimated density $\hat{f}_{5.94}$
of the points $\{\hat{p}_i(5.94)\}_{i=1,\ldots,215}$ (and it will be discussed later). It appears
clearly that the behaviour of these probability curves is quite heterogeneous.

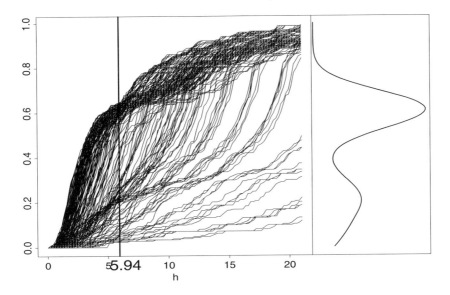

Fig. 9.3. Spectrometric Data: the Probability Curves

More generally, we propose to use such probability curves both for building a partition of our sample and choosing a bandwidth h. In fact, we are interested by the bandwidth which reveals the largest heterogeneity of the considered sample, which is a way to define homogeneous classes and thus a partition. More precisely, for a fixed bandwidth h, we have at hand n *probability points* $\{\widehat{p}_i(h)\}_{i=1,\ldots,n}$ for which it is easy to estimate its real density $f_h(.)$ via any standard estimator.

Splitting the probability points and partitioning the sample. Instead of splitting the functional observations $\chi_1, \ldots, \chi_n$ themselves, we propose to split the set of points $\{\widehat{p}_i(h)\}_{i=1,\ldots,n}$. In particular, partitioning is all the easier as $f_h(.)$ has several modes. Indeed, suppose that $f_h(.)$ has K modes and denote by $f_h(m_{1,\mathcal{S}}), \ldots f_h(m_{K-1,\mathcal{S}})$ the corresponding $K-1$ local minima of $f_h(.)$. Assume that $m_{1,\mathcal{S}} < \cdots < m_{K,\mathcal{S}}$. Then, we can build the following subsets for $j = 1, \ldots, K$:

$$\mathcal{I}_j = \{i \in \{1, \ldots, n\}, \ m_{j-1,\mathcal{S}} < \widehat{p}_i(h) \le m_{j,\mathcal{S}}\},$$

where $m_{0,\mathcal{S}} = 0$ and $m_{K,\mathcal{S}} = 1$. In other words, $\mathcal{I}_j$ is the set of the i's for which the corresponding probability points admit $M_{k,\mathcal{S}}$ as the closest mode and we have:

$$\{1, \ldots, n\} = \cup_{k=1}^{K} \mathcal{I}_k \text{ and } \forall k \neq k', \ \mathcal{I}_k \cap \mathcal{I}_{k'} = \phi.$$

Now, it is clear that we get a partition of the sample $\mathcal{S}$ just by defining:

$$\mathcal{S}_k = \{\chi_i, i \in \mathcal{I}_k\}, k = 1, \ldots, K.$$

Selecting the bandwidth h. This the reason why we would like that $f_h(.)$ has several modes and why we select a bandwidth h for which the set $\{\widehat{p}_i(h)\}_{i=1,\ldots,n}$ reveals the largest heterogeneity (i.e., several modes) as possible. This can be done by minimizing the entropy of $f_h(.)$ over a set of bandwidths H:

$$h_{opt} = \arg \inf_{h \in H} \int_0^1 f_h(t) \log f_h(t) \, dt,$$

the support of $f_h(.)$ being $(0,1)$ since the probability points belong to $[0,1]$. It is well known that the smaller the entropy the larger the heterogeneity of the set $\{\widehat{p}_i(h)\}_{i=1,\ldots,n}$ (i.e., we can expect that $f_h(.)$ has several modes when its entropy is small). From a practical point of view, it suffices to replace $f_h(.)$ with any standard density estimator $\widehat{f}_h(.)$ and H with a suitable finite set $\widehat{H}$. Finally, we are able to compute

$$\widehat{h}_{opt} = \arg \min_{h \in \widehat{H}} \int_0^1 \widehat{f}_h(t) \log \widehat{f}_h(t) \, dt.$$

Figure 9.3 points out the resulting bandwidth of such an automatic procedure ($\widehat{h}_{opt} = 5.94$) for the spectrometric curves whereas the right part displays $\widehat{f}_{5.94}$. In this application, we have used the familiar Parzen-Rosenblatt unidimensional kernel density estimate:

$$\widehat{f}_h(t) = \frac{1}{nb} \sum_{i=1}^{n} K_0 \left(b^{-1}(t - \widehat{p}_i(h)) \right),$$

K_0 being a standard symmetrical kernel (for instance a kernel of type 0). Of course, it could be the case on other situations that one gets a unimodal structure for the estimated density $\widehat{f}_{\widehat{h}_{opt}}$, and in this case the sample would not have to be split. Moreover, if the partitioning is possible, we perform it only if the splitting score is greater than the fixed threshold. So, we have at hand two ways to stop the splitting procedure.

9.4.3 Classification Algorithm

It is time to describe the iterative classification algorithm which is a summary of the previously detailed stages:

<u>STEP 1</u> Computing $\widehat{h}_{opt}$:
      if $\widehat{f}_{\widehat{h}_{opt}}$ admits several modes
        then goto <u>STEP 2</u>

        else goto STEP 3

STEP 2 Splitting the sample and compute SC:
        if $SC > \tau$
           then achieve STEP 1 for each subclass
           else goto STEP 3

STEP 3 Stop.

As one can see, this algorithm is very simple as soon as we are able to compute $\widehat{h}_{opt}$ and $\widehat{f}_{\widehat{h}_{opt}}$ for any sample. Note that, as with any classification procedure, the final partition obtained by this algorithm is very sensitive to the choice of the threshold τ. In addition, another source of variability of the method comes from the random subsampling step used to compute the heterogeneity index SHI.

9.4.4 Implementation: R/S+ Routines

Because such analyses use a recursive algorithm, the programming is more complicated than for previous problems (prediction and discrimination). We refer the reader to the companion website[1] for loading routines, detailed descriptions of the data as well as commandlines for achieving case studies and plots. The automatic routine for classifying functional data is called `classif.automatic`. As in any other routine described earlier the user has to fix a semi-metric and a kernel function. Moreover, the practitioner may control the threshold splitting score τ as well as the number of subsamples involved in the computation of the heterogeneity index SHI.

9.5 Nonparametric Unsupervised Classification in Action

We propose to classify the spectrometric curves described in Section 7.2.1 where the response variable (i.e., Col 101) is ignored in order to respect the unsupervised setting. According to our experience (see Sections 7.2 and 8.4.2), we know that the semi-metric based on the second derivative is well adapted for these data. So, we keep this semi-metric, recalling that it is defined by

$$d_2^{deriv}(\chi, \xi) = \sqrt{\int \left(\chi^{(2)}(t) - \xi^{(2)}(t) \right)^2 dt}$$

and refering to Section 3.4.3 for more details. Of course, the reader can select another semi-metric and see the behaviour of the classification procedure. In addition, the following heterogeneity index is considered:

----

[1] *http://www.lsp.ups-tlse.fr/staph/npfda*

$$HI(\mathcal{S}) \;=\; \frac{d_2^{deriv}(\boldsymbol{\mathcal{X}}_{mod,\mathcal{S}}, \boldsymbol{\mathcal{X}}_{mean,\mathcal{S}})}{d_2^{deriv}(\boldsymbol{\mathcal{X}}_{mod,\mathcal{S}}, 0) + d_2^{deriv}(\boldsymbol{\mathcal{X}}_{mean,\mathcal{S}}, 0)}.$$

Of course, this heterogeneity index could have been defined by replacing the mean with the median. Here, the functional mean is used because it makes sense for such a dataset. (See discussion in Section 9.2.1). Figure 9.4 draws our classification tree and precises the splitting scores for each partitioning and the subsampling heterogeneity indices for each group. Figure 9.5 displays the spectrompetric curves corresponding to the three terminal leaves of our classification tree (GROUP 1, GROUP 21 and GROUP 22). Concerning the size of the classes, GROUP 1 (resp. GROUP 21 and GROUP 22) contains 135 (resp. 30 and 50) curves.

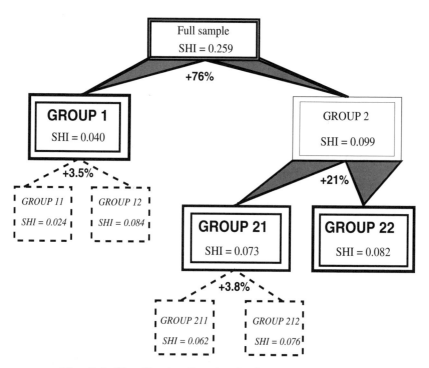

Fig. 9.4. Classification Tree for the Spectrometric Curves

It is clear that the selected semi-metric emphasizes features of the second derivatives instead of the spectrometric curves themselves. The differences between groups come essentially from the shape of the second derivatives at the two valleys around the wavelengths 930 and 960 and the two peaks around 910 and 950. The classification of the spectrometric curves is drawn by the amplitude of their second derivatives at these wavelengths. Note that our bandwidth choice procedure allows us to give a good estimation of the functional modes in the sense that they appear as a good summary of each computed class. In addition, Figure 9.6 focuses on the behaviour of the splitting score along the classification procedure. This last plot is very helpful for stopping the splitting procedure. Indeed, it appears that the gain that one would obtain by splitting GROUP 1 (resp. GROUP 21) would be 3.5% (resp. 3.8%) and these small gains motivate that these two groups appear as terminal leaves of the classification tree. Note that GROUP 22 is not split because the corresponding estimated density $\widehat{f}_{\widehat{h}_{opt}}$ turns out to be unimodal.

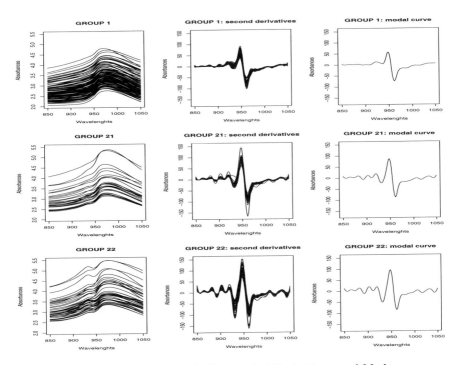

Fig. 9.5. Spectrometric Curves, 2nd Derivatives, and Modes

9.6 Theoretical Advances on the Functional Mode

As explained at the begining of this chapter, we focus here only on the estimation of the mode θ of the probability distribution of a functional random

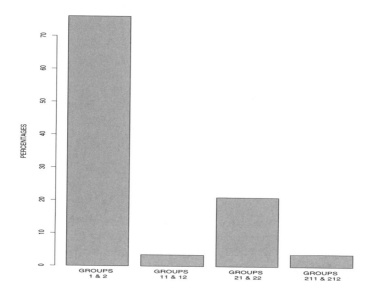

Fig. 9.6. Behaviour of the Splitting Score for the Spectrometric Data

variable $\mathcal{X}$. The reason comes from the fact that the mode appears as a central notion of our functional classification method. We recall that we define an estimator θ_n of the mode θ as the supremum of a pseudo kernel-type estimator $\tilde{f}$ of the density f of $\mathcal{X}$ with respect to a measure μ σ-finite, diffuse and such that $0 < \mu(B) < \infty$ for all open ball B of E. The theoretical advances in this field are quite developed. Therefore we will state only the almost complete convergence (without rate) of θ_n. To this end, we will first show asymptotic results for $\tilde{f}$ in a uniform way. Because of the uniformity aspect, we have to introduce an assumption driving the uniform behaviour of the probability distribution of the f.r.v. $\mathcal{X}$.

9.6.1 Hypotheses on the Distribution

The main difference with asymptotics developed in the previous chapters comes from the fact that we have to control in some uniform way the behaviour of the small ball probability function:

$$\varphi_{\chi}(.) = P(\mathcal{X} \in B(\chi, .)).$$

To do that, we introduce a strictly positive and increasing function $\psi(.)$ which does not depend on χ and will play a similar role as $\varphi_{\chi}(.)$. In order to take into account the uniformity aspect, we suppose that

$$\begin{cases} \lim_{t \to 0} \sup_{\chi \in \mathcal{C}} \left| \frac{P(X \in B(\chi, t))}{\psi(t)} - f(\chi) \right| = 0, \\ \qquad \text{with} \\ \lim_{t \to 0} \psi(t) = 0 \text{ and } \exists C > 0, \exists \epsilon_0, \forall \epsilon < \epsilon_0, \int_0^\epsilon \psi(u) du > C\epsilon\psi(\epsilon), \end{cases} \qquad (9.1)$$

where $\mathcal{C}$ is a subset of E and $B(\chi, t)$ is the ball of center χ and radius t. This is clearly a hypothesis acting on the shape of small ball probabilities which is more restrictive than Hypothesis (6.2) introduced in Section 6.2.1. Following the ideas developed by [ABC03], let us consider the subset

$$\mathcal{C}_\epsilon = \{\chi \in \mathcal{C}, \ f(\theta) - f(\chi) < \epsilon\},$$

where θ denotes the mode which is defined as follows:

$$\theta = \arg \sup_{\chi \in \mathcal{C}} f(\chi).$$

Then, we suppose that f satisfies the following conditions:

$$\lim_{\epsilon \to 0} Diam(\mathcal{C}_\epsilon) = 0. \qquad (9.2)$$

It is worth noting that Hypothesis (9.1) is a mixture of conditions on the size of the subset $\mathcal{C}$, on the behaviour of small ball probability and on the smoothness property of f. In order to come back to a formulation which is close to the standard condition made in finite dimensional setting (see Section 9.7.4 for references), let us look at the following special case.

Remark 9.3. If f is uniformly continuous on $\mathcal{C}$ and μ is asymptotically invariable on $\mathcal{C}$ in the sense that for t small enough

$$\forall x, \mu(B(\chi, t)) = \mu(B(0, t))(1 + o(1)),$$

then (9.1) is checked with $\psi(t) = \mu(B(0, t))$.

Indeed, it suffices to write that

$$P(B(\chi, t)) = f(\chi)\mu(B(\chi, t)) + \int_{B(\chi, t)} (f(\zeta) - f(\chi)) \, d\mu(\zeta),$$

which allows us to get for t small enough that

$$\frac{P(B(\chi, t))}{\mu(B(0, t))} - f(\chi) \sim \frac{\int_{B(\chi, t)} (f(\zeta) - f(\chi)) \, d\mu(\zeta)}{\mu(B(0, t))}.$$

Because of the uniform continuity of f on $\mathcal{C}$, it holds that

$$\forall \epsilon > 0, \exists \eta_\epsilon, \forall \chi \in \mathcal{C}, d(\chi, \zeta) < \eta_\epsilon, \ |f(\chi) - f(\zeta)| < \epsilon,$$

which implies that for t small enough:

$$\forall \chi \in \mathcal{C}, \left| \frac{P(B(\chi,t))}{\mu(B(0,t))} - f(\chi) \right| \leq \frac{\mu(B(\chi,t))}{\mu(B(0,t))} \epsilon,$$

which achieves the proof of this remark. $\square$

Concerning Hypothesis (9.2), to fix the ideas we come back to the one-dimensional case by recalling a result coming from [Ca02]. In fact, in the one-dimensional setting, this hypothesis appears clearly as a flatness condition on f around θ (see references in Section 9.7.4).

Remark 9.4. If $E = \mathbb{R}$ and $d(.,.)$ is the usual metric in $\mathbb{R}$, if f is k-times $(k \geq 2)$ continuously differentiable around θ with $f^{(j)}(\theta) = 0$, $\forall j = 1, \ldots, k-1$ and $|f^{(k)}(\theta)| > 0$, then $Diam(\mathcal{C}_\epsilon) = O\left(\epsilon^{1/k}\right)$.

It suffices to write the Taylor expansion of f

$$f(x) - f(\theta) = (x - \theta)^k (C + o(1)),$$

where C does not depend on x. Thus, we have

$$\forall (x, x') \in \mathcal{C}_\epsilon, |x - x'| \leq |x - \theta| + |\theta - x'|,$$

which implies that

$$Diam(\mathcal{C}_\epsilon) = O\left(\epsilon^{1/k}\right).$$

This achieves the proof of this remark. $\square$

9.7 The Kernel Functional Mode Estimator

9.7.1 Construction of the Estimates

Let us now state precisely the pseudo-estimator $\tilde{f}$ of f as follows:

$$\forall \chi \in E, \ \tilde{f}(\chi) = \frac{1}{n\,Q(K,h)} \sum_{i=1}^{n} K\left(h^{-1}d(\chi, \boldsymbol{\mathcal{X}}_i)\right),$$

where

$$Q(K, h) = -\int_0^1 K'(t)\psi(ht)\,dt,$$

K being a kernel function. This is not exactly an estimator because the constant of normalization $Q(K,h)$ is unknown but it is important to remark that it does not depend on χ. Thus, for estimating the mode, it comes to the same thing to maximize $\tilde{f}(.)$ or $\sum_{i=1}^{n} K\left(h^{-1}d(., \boldsymbol{\mathcal{X}}_i)\right)$. Finally, the estimator θ_n of the mode θ is defined as:

$$\theta_n = \arg\sup_{\chi \in \mathcal{C}} \tilde{f}(\chi) = \arg\sup_{\chi \in \mathcal{C}} \sum_{i=1}^{n} K\left(h^{-1}d(\chi, \mathbfcal{X}_i)\right).$$

In order to control the size of the set in which we look for the mode, we suppose that it verifies the following assumption:

$$\begin{cases} \exists \alpha > 0, \beta > 0, \ \mathcal{C} \subset \cup_{k=1}^{d_n} B(c_k, r_n) \\ \text{with } d_n = n^\alpha \text{ and } r_n = o\left(h\,\psi(h)^{1/\beta}\right). \end{cases} \tag{9.3}$$

We first study the uniform almost complete convergence of $\tilde{f}$. To do that, the estimator has to check the following conditions:

$$\begin{cases} h \text{ is a positive sequence such that} \\ \lim_{n\to\infty} h = 0 \text{ and } \lim_{n\to\infty} \dfrac{\log n}{n\,\psi(h)} = 0, \end{cases} \tag{9.4}$$

and

$$\begin{cases} K \text{ is a kernel of type II such that} \\ \exists C < \infty, \ \forall(x, x') \in \mathbb{R}^2, \ |K(x) - K(x')| \le C|x - x'|^\beta. \end{cases} \tag{9.5}$$

The conditions acting on h are standard except that $\psi(h)$ plays the role of the usual small ball probability function $\varphi_\chi(h)$ (see Hypothesis (6.3)). As is usual in the multivariate case, the Lipschitz-type condition for the kernel K is necessary for obtaining asymptotical results in a uniform way. This is the reason why Hypothesis (9.5) is more restrictive than Hypothesis (6.3) because in particular the kernels of type I have to be excluded. We are ready for giving the main results.

9.7.2 Density Pseudo-Estimator: a.co. Convergence

Theorem 9.5. *Under the probability condition (9.1), if $\mathcal{C}$ satisfies (9.3) and if the pseudo kernel-type estimator verifies (9.4) and (9.5), then we have:*

$$\lim_{n\to\infty} \sup_{\chi \in \mathcal{C}} \left|\tilde{f}(\chi) - f(\chi)\right| = 0, \ a.co. \tag{9.6}$$

Proof. First, let us remark that

$$\sup_{\chi \in \mathcal{C}} \left|\tilde{f}(\chi) - f(\chi)\right| \le \sup_{\chi \in \mathcal{C}} \left|\tilde{f}(\chi) - \mathbb{E}\tilde{f}(\chi)\right| + \sup_{\chi \in \mathcal{C}} \left|\mathbb{E}\tilde{f}(\chi) - f(\chi)\right|.$$

So, Theorem 9.5 will be proved as soon as Lemmas 9.6 and 9.7 below will be stated.□

Lemma 9.6. *Under conditions of Theorem 9.5, we have*

$$\lim_{n\to\infty} \sup_{\chi\in\mathcal{C}} \left| \mathbb{E}\tilde{f}(\chi) - f(\chi) \right| = 0.$$

Proof. It is easy to see that

$$f(\chi) - \mathbb{E}\tilde{f}(\chi) = f(\chi)\, R_n(\chi), \tag{9.7}$$

where

$$R_n(\chi) = \frac{Q(K,h)f(\chi) - \mathbb{E}K\left(h^{-1}d(\chi,\boldsymbol{\mathcal{X}})\right)}{Q(K,h)f(\chi)}.$$

Now, let us focus on the quantity $R_n(\chi)$. According to the proof of Lemma 4.4 (see (4.8)), condition (9.5) implies that

$$\mathbb{E}K\left(h^{-1}d(\chi,\boldsymbol{\mathcal{X}})\right) = -\int_0^1 K'(u)\left(\frac{P\left(B(\chi,hu)\right)}{\psi(hu)} - f(\chi)\right)\psi(hu)\,du$$
$$+ Q(K,h)\,f(\chi). \tag{9.8}$$

Then, we have

$$f(\chi)\,|R_n(\chi)| \le -\frac{1}{Q(K,h)}\int_0^1 K'(u)\left|\frac{P\left(B(\chi,hu)\right)}{\psi(hu)} - f(\chi)\right|\psi(hu)\,du.$$

On the other hand, from (9.7) we deduce

$$\sup_{\chi\in\mathcal{C}}\left|\mathbb{E}\tilde{f}(\chi) - f(\chi)\right| \le \sup_{\chi\in\mathcal{C}} f(\chi)\,|R_n(\chi)|,$$

and (9.1) leads us to the claimed result. $\square$

Lemma 9.7. *Under conditions of Theorem 9.5, we have*

$$\lim_{n\to\infty}\sup_{\chi\in\mathcal{C}}|\tilde{f}(\chi) - E\tilde{f}(\chi)| = 0, \ a.co.$$

Proof. Because $\mathcal{C} \subset \cup_{k=1}^{d_n} B(c_k, r_n)$, we have

$$P\left(\sup_{\chi\in\mathcal{C}}|\tilde{f}(\chi) - \mathbb{E}\tilde{f}(\chi)| > \eta\right) \le \underbrace{P\left(\sup_{\chi\in\mathcal{C}}|\mathbb{E}\tilde{f}(\chi) - \mathbb{E}\tilde{f}(c_{k(\chi)})| > \frac{\eta}{3}\right)}_{T_1}$$
$$+ \underbrace{P\left(\sup_{\chi\in\mathcal{C}}|\tilde{f}(\chi) - \tilde{f}(c_{k(\chi)})| > \frac{\eta}{3}\right)}_{T_2}$$
$$+ \underbrace{d_n \max_{l=1,\dots,d_n} P\left(|\tilde{f}(c_l) - \mathbb{E}\tilde{f}(c_l)| > \frac{\eta}{3}\right)}_{T_3},$$

where $c_{k(\chi)}$ denotes the closest center to χ among $\{c_1,\dots,c_{d_n}\}$. To simplify notations, let $K_{i,\chi}$ denotes $K\left(h^{-1}d(\chi,\boldsymbol{\mathcal{X}}_i)\right)$ for any $\chi \in E$.

- Concerning T_1 and T_2:

$$\left| \tilde{f}(\chi) - \tilde{f}(c_{k(\chi)}) \right| \leq \frac{1}{n\,Q(K,h)} \sum_{i=1}^{n} \left| K_{i,\chi} - K_{i,c_{k(\chi)}} \right|$$

$$\leq \frac{C\,r_n^{\beta}}{h^{\beta}\,Q(K,h)},$$

the last inequality coming from (9.5). Now, it suffices to remark that the second part of (9.1) implies that

$$1/Q(K,h) = O(1/\psi(h)). \tag{9.9}$$

Now just combine (9.3) with (9.9) in order to state that for n large enough, $T_1 = T_2 = 0$.

- Concerning T_3: it is easy to see that

$$\tilde{f}(c_l) - \mathbb{E}\tilde{f}(c_l) \;=\; \frac{1}{n} \sum_{i=1}^{n} Z_{i,c_l,n},$$

with $Z_{i,c_l,n} = (K_{i,c_l} - \mathbb{E}K_{1,c_l})/Q(K,h)$. Using (9.5) and (9.9) allows us to write

$$|Z_{i,c_l,n}| \leq C/\psi(h).$$

Let us now focus on the variance of $Z_{i,c_l,n}$:

$$Var\left(Z_{i,c_l,n}\right) \;=\; \left(\mathbb{E}K_{i,c_l}^2 - (\mathbb{E}K_{1,c_l})^2\right)/Q(K,h)^2.$$

Taking $\mathcal{K}(.) = K^2(.)/C_K$ with $C_K = \int_0^1 K^2(u)\,du$, the last equation becomes

$$Var\left(Z_{1,c_l,n}\right) \;=\; \frac{C_K^2}{Q(K,h)^2}\left(\mathbb{E}\mathcal{K}_{1,c_l} - (\mathbb{E}\mathcal{K}_{1,c_l})^2\right).$$

Because $\mathcal{K}$ is also a kernel of type II, we can apply (9.8) with $\mathcal{K}$ and it comes out that

$$\mathbb{E}\mathcal{K}_{1,c_l} \;\leq\; C_1\,Q(\mathcal{K},h).$$

In addition, we can check easily that

$$Q(\mathcal{K},h) \;\leq\; C_2\,Q(K,h).$$

On the other hand, because of (9.8) we have

$$\frac{\mathbb{E}K_{1,c_l}}{Q(K,h)} < C_3,$$

and using (9.9), we have

$$Var\left(Z_{1,c_l,n}\right) \;\leq\; C_4/\psi(h).$$

Then by using Bernstein-type exponential inequality (see Corollary A.9-ii) we get

$$T_3 \leq C\, n^\alpha \, \exp\left\{-C'\,\eta^2\, n\,\psi(h)\right\}$$

$$\leq C\, n^\alpha \, \exp\left\{-C'\,\eta^2\, \log n\, \left(\frac{n\,\psi(h)}{\log n}\right)\right\}.$$

Because of (9.4), we have

$$\forall b > 0, \exists n_b, \forall n > n_b, \ n\,\psi(h)/\log n > b,$$

which implies that

$$T_3 \leq C\, n^{\alpha - C'\, b\, \eta^2}.$$

Then, $\forall \eta > 0$, we can take b large enough such that $\alpha - C'\, b\, \eta^2 < -1$, which implies that $\lim_{n \to 0} T_3 = 0$, $a.co.$ $\square$

9.7.3 Mode Estimator: a.co. Convergence

Once the asymptotic property stated for the kernel pseudo-estimator of the density, we are ready to give the almost complete convergence of the estimator θ_n of the mode θ.

Theorem 9.8. *Under the conditions (9.1), (9.2), (9.3), (9.4) and (9.5), we have:*

$$\lim_{n \to \infty} d(\theta, \theta_n) = 0, \quad a.co. \tag{9.10}$$

Proof. Because of the condition (9.2), we have:

$$\forall \eta > 0, \ \exists \epsilon > 0, \ \forall \chi \in \mathcal{C}, \ d(\theta, \chi) > \eta \ \Rightarrow \ |f(\theta) - f(\chi)| > \epsilon.$$

In particular, by taking $\chi = \theta_n$, we have

$$P\left(d(\theta, \theta_n) > \eta\right) \leq P\left(|f(\theta) - f(\theta_n)| > \epsilon\right).$$

So, Theorem 9.8 will be proved as soon as we will be able to show that

$$\lim_{n \to \infty} f(\theta_n) = f(\theta), \quad a.co.$$

To do that, let us remark that

$$|f(\theta) - f(\theta_n)| \leq \left|f(\theta) - \tilde{f}(\theta_n)\right| + \left|\tilde{f}(\theta_n) - f(\theta_n)\right|,$$

$$\leq \left|\sup_{\chi \in \mathcal{C}} f(\chi) - \sup_{\chi \in \mathcal{C}} \tilde{f}(\chi)\right| + \sup_{\chi \in \mathcal{C}}\left|\tilde{f}(\chi) - f(\chi)\right|,$$

$$\leq 2 \sup_{\chi \in \mathcal{C}}\left|\tilde{f}(\chi) - f(\chi)\right|.$$

It suffices now to apply Theorem 9.5 in order to achieve the proof. $\square$

9.7.4 Comments and Bibliography

Main differences with the other results presented in this book come from one side from the uniform convergence needed for the density estimator of a functional random variable (see Theorem 9.5), and from the other side from the existence of the density function itself. Because of these technical difficulties, the results are quite less developed than in other parts of the book and many interesting open problems remain to be studied. In addition we will see in Part III of this book that while all other functional methods presented before can be extended to non-necessarily independent functional samples, such an extension is still unknown in functional mode estimation setting. We will just discuss here two interesting open questions which are really specific to the functional mode estimation setting and for which we can propose some track for answering.

> **Open question 7: Rates of convergence.** *Is it possible to get rates of convergence?*

As discussed in Remark 9.3, Hypothesis (9.1) is linked with a continuity-type functional nonparametric model, and it is in accordance with the remainder of this book to obtain only the almost complete convergence (without rates). In return, rates of convergence could be reached by imposing more restrictive smoothness conditions on the density operator f. Indeed, a first approach could consist in introducing a Lipschitz-type nonparametric model for the density f and in combining it with a new small ball probability condition such as for some $\gamma > 0$:

$$\limsup_{t \to 0} \sup_{\chi \in \mathcal{C}} \left| \frac{P(X \in B(\chi, t))}{\psi(t)} - f(\chi) \right| = O(t^\gamma),$$

in other to get rates of convergence. Another way could be to consider more restrictive regularity properties on the density (such as differentiability for instance) but from a technical point of view this could be more difficult than in the multivariate setting (see however [DFV06] for first advances in this direction).

> **Open question 8: Extension to other sets.** *Is it possible to get convergence over wider sets $\mathcal{C}$? over random sets?*

To see the problem concerning the first question, let us focus on the technical condition (9.3) acting on the set $\mathcal{C}$. This hypothesis is crucial for controling the size of $\mathcal{C}$ in order to get uniform convergence of the density estimate itself. Therefore, the asymptotic result given in Theorem 9.5 is proved only on such sets. One track for a further theoretical investigation would be to extend the results to more general sets because the size of $\mathcal{C}$ depends on the knowledge we have a priori on the location of the mode, since by construction the mode is assumed to belong to such a set. In order to give an idea with regard to the second part of this question, recall that in practice a simple mode estimate is obtained by maximizing the density over the sample (see Section 9.2.3) and so over a random set. One possible way to solve this problem could follow the idea in [ABC03]. Intermediate approaches toward this question would be to look for the mode in a *smooth* functional space like Sobolev spaces or to regularize directly the simple mode estimation obtained over the sample.

To conclude these comments, we will give now a short overview on the literature from the finite dimensional case to the infinite one. Starting with [P62], kernel mode estimation has been widely studied (key references are including [E80], [E82] and [V96]) while most recent advances are in [ABC03], [ABC04] and [HZ04]. Theorem 9.8 can be seen as an infinite dimensional version of some of the results presented in these references, but of course the state of the art in finite dimensional setting is quite a bit more developed. In particular, the links between the smoothing parameter (i.e., the bandwidth) and possible multimodalities have been widely studied (see for instance [S92], [M95] and [JMVS95]). Such studies remain a challenge for infinite dimensional variables. In fact, for functional variables, few authors have studied the mode because of the under-development of the statistical knowledge on the density estimation itself. Apart from an earlier work by [G74], the literature on this last topic seems limited to the recent contributions by [JO97], [D04] and [D04b] (see also [D02] for more discussion). So, the amount of statistical works about the infinite dimensional mode is even more restricted since as far as we know it just involves [DFV06] for theoretical advances and [GHP98] and [DFV04] for practical ones.

9.8 Conclusions

From a practical point of view, the functional classification methodology provides good results with respect to our spectrometric dataset. Of course, it can be improved by producing other more sophisticated semi-metrics or estimations of various features of the functional distribution. With regard to the semi-metric choice problem, the building of adaptive indices of proximity is a general and crucial problem which concerns all the methods developed in this book (see the general discussion in Chapter 13).

It is worth noting that there exist various ways to classify curves but almost all the methods in the literature come down to standard multivariate techniques. For instance, one can think to classify directly the discretized curves or their coordinates in some basis expansion. In this way, all standard classification methods are usable in the functional data setting (see for instance the monographs [Go99] and [H97] and the references therein). In particular, other clustering algorithms such as the k-means one (see *e.g.* [H75] and [HW79]) can be used (see [ACMM03] and [TK03] for examples of the k-means algorithm acting on respectively B-spline Fourier expansions of the curves).

Alternatively, in this chapter we presented "pure" functional nonparametric methods in the sense that we always developed tools especially for taking into account functional aspects. This unsupervised classification technique completes the standard curves classification methods. Of course, as it has been pointed out throughout this chapter, we are fully aware that this chapter is just a starting point for a new field of functional statistics in which many aspects deserve further investigation. In particular, one might be interested in the following question:

> **Open question 9 :   What about a fully functional k-means procedure?**

The proximity notions discussed in Chapter 13 could be a first step in this direction. In this spirit, one can mention the recent work by [CF06] which uses proximities based on the trimmed mean notion for classifying functional data through k-means algorithm.

Part IV

Nonparametric Methods for Dependent
Functional Data

An important field of application of functional statistical methods concerns the analysis of continuous time stochastic processes. Splitting such a process in several different periods of time allows us to construct a functional data set: each period of the process being considered as a functional data. The main feature of such type of dataset is linked with the dependence structure existing between the statistical units (that is, between the periods). This prevents us from using directly any satistical method developed for i.i.d. samples in such a situation. So there is a real interest in looking at how each nonparametric functional method developed earlier in this book is behaving, both theoretically and from a practical point of view, for dependent data. Because the guideline of this book is concerning nonparametric models, we focus on well-adapted dependence structures based on mixing type conditions (alternative functional parametric dependence structures can be found in [B00]).

This part is organized as follows: In Chapter 10 we will discuss some dependence structure based on mixing processes modelizations, for which the probabilistic backround is sufficiently developed to allow for theoretical advances on nonparametric dependent functional data analysis. Not to mask our main purpose, this probabilistic backround will not be presented in the main body of the book but quickly recalled and discussed in Appendix A. In Chapter 11 most of the theoretical results presented earlier for i.i.d. functional variables are extended to mixing ones. From one part, these extensions are concerning the prediction results of Part II, and asymptotic results for mixing variables will be given about kernel estimates of regression, conditional density and conditional c.d.f. (including their applications to functional conditional mean, median, mode and quantiles). From a second part, these extensions will concern the kernel based curves classification method discussed along Part III. Finally, we will return in Chapter 12 to continuous time processes. It will be discussed how these advances on functional nonparametric dependent statistics can be applied in forecasting future values of some time series. Through a quick application to the economic dataset presented before in Section 2.3, it will also be seen in Chapter 12 how the R and S+ routines presented before for i.i.d. functional datasets can be useful in this new time series context.

Our main goal in this part is to show how the dependence is acting on the behaviour of nonparametric functional methods. Therefore, not to mask this main objective, we have decided to present all the following chapters in a synthetic way which emphasizes what is changing in the mixing situation in comparison with the i.i.d. case. In other words, everything which behaves as in the independent setting will not be discussed in detail but we will make abundant references to previous parts of this book. This will concern the theoretical advances presented in Chapter 11, as well as the applied issues and the R and S+ routines presented in Chapter 12. Of course, such synthetic presentations have the drawback of obliging the reader to keep in mind previous chapters of our book. However, it has the great advantage of highlighting the influence of the dependence structure, and also to avoid unuseful repetition.

10

Mixing, Nonparametric and Functional Statistics

Modelling dependence is of great interest in statistics. This comes mainly from the fact that it opens the door for application involving time series. Naturally, this question should be attacked in the nonparametric functional data context of this book. These statistical motivations will not be discussed in this chapter but later on in Chapter 12. The aim of this chapter is to recall some basic definitions, to discuss briefly the existing literature in finite dimensional setting and to introduce notations and general advances in the functional setting. This will prepare the way for the theoretical advances in nonparametric functional statistics for dependent samples that will be presented in Chapter 11.

10.1 Mixing: a Short Introduction

Mixing conditions are usual structures for modelling dependence for a sequence of random variables. It is out of our scope to present an exhaustive discussion on this point. A good overview of probabilistic knowledges on this notion can be found for instance in [Y92] (other references could be [B86] or [D95]). The reader may be interested also in the monographs by [GHSV89], [Y94], or [B98] which are centered both on the mixing structures themselves and on their interest for nonparametric statistics. The monographs [Y93a] and [Y93b] discuss the interest of mixing for statistical settings other than nonparametric estimation. Before going into the use of mixing notions in nonparametric statistics let us first recall some definitions and fix some notations. In our book we focus on the α-mixing (or strong mixing) notion, which is one of the most general among the different mixing structures introduced in the literature (see for instance [RI87] or Chapter 1 in [Y94] for definitions of various other mixing structures and links between them). This strong mixing notion is defined in the following way.

Let $(\xi_n)_{n \in \mathbb{Z}}$ be a sequence of random variables defined on some probabilistic space $(\Omega, \mathcal{A}, P)$ and taking values in some space $(\Omega', \mathcal{A}')$. Let us denote, for

$-\infty \leq j \leq k \leq +\infty$, by $\mathcal{A}_j^k$ the σ-algebra generated by the random variables $(\xi_s, j \leq s \leq k)$. The strong mixing coefficients are defined to be the following quantities:

$$\alpha(n) = \sup_{k} \sup_{A \in \mathcal{A}_{-\infty}^{k}} \sup_{B \in \mathcal{A}_{n+k}^{+\infty}} |P(A \cap B) - P(A)P(B)|.$$

The following definition of mixing processes was originally introduced by [R56].

Definition 10.1. *The sequence $(\xi_n)_{n \in \mathbb{Z}}$ is said to be α-mixing (or strongly mixing), if*

$$\lim_{n \to \infty} \alpha(n) = 0.$$

In the remainder of this book, in order to simplify the presentation of the results and not to mask our main purpose, we will mainly consider both of the following subclasses of mixing sequences:

Definition 10.2. *The sequence $(\xi_n)_{n \in \mathbb{Z}}$ is said to be arithmetically (or equivalently algebraically) α-mixing with rate $a > 0$ if*

$$\exists C > 0, \ \alpha(n) \leq C n^{-a}.$$

It is called geometrically α-mixing if

$$\exists C > 0, \ \exists t \in (0,1), \ \alpha(n) \leq C t^n.$$

10.2 The Finite-Dimensional Setting: a Short Overview

Since a long time ago, and starting with earlier advances provided for instance by [R69] or [Ro69], mixing assumptions have been widely used in nonparametric statistics involving finite dimensional random variables. It is really out of purpose to make here a presentation of the very plentiful literature existing in this field, but one could reasonably say that almost all the results stated in finite dimensional nonparametric statistics for i.i.d. variables have been extended to mixing data. Key previous papers in this direction were those concerning regression estimation from mixing samples (see for instance [R83] and [C84]), but now extensions of nonparametric methods for mixing variables are available in most problems involving density, hazard, conditional density, conditional c.d.f., spectral density, etc. The reader who would be interested in having a good overview of all the literature should look at some among the following (un-exhaustive) list of works [GHSV89], [T94], [Y94], [B98] or [FY04].

We will see in Chapter 11 that these general ideas remain true in the setting of functional dependent samples, since almost all the results stated in previous parts of this book will be extended to strong mixing infinite dimensional variables.

10.3 Mixing in Functional Context: Some General Considerations

Even if the above-discussed nonparametric literature quite often concerns only finite dimensional variables, it is worth being noted that the general definitions presented in Section 10.1 are available for infinite dimensional variables. For the purpose of this book we have to deal with random variables taking values in some semi-metric space. The aim of this section is to present some general results that will be used throughout this book (this is more specifically true for Proposition 10.4). These results allow us (at least in our kernel framework) to treat mixing sequences of f.r.v. by using (mainly) probabilistic results for mixing sequences of r.r.v. Because these results could be useful for other purposes than our kernel nonparametric framework and because we did not find them published elsewhere (at least in a simple enough form directly applicable for statistical purposes), we have decided to spend some space to present short proofs of them even if they are clearly going beyond the scope of our book. The first result concerns the case when Ω' is a semi-normed space whereas the second one is an extension to semi-metric space. Note that the second proposition will be intensively used throughout the remainder of the book.

Proposition 10.3. *Assume that Ω' is a semi-normed space with semi-norm $||.||$, and that $\mathcal{A}'$ is the σ-algebra spanned by the open balls for this semi-norm. Then we have:*

i) $(\xi_n)_{n \in \mathbb{Z}}$ is α-mixing $\Rightarrow (||\xi_n||)_{n \in \mathbb{Z}}$ is α-mixing;

ii) In addition, if the coefficients of $(\xi_n)_{n \in \mathbb{Z}}$ are geometric (resp. arithmetic) then those of $(||\xi_n||)_{n \in \mathbb{Z}}$ are also geometric (resp. arithmetic with the same order).

Proof. Let $\xi = (\xi_{j_1}, \dots \xi_{j_l})$, with $-\infty \le j_1 \le \dots \le j_l \le +\infty$, be some fixed subfamily of $(\xi_n)_{n \in \mathbb{Z}}$. Denote by $||\xi|| = (||\xi_{j_1}||, \dots ||\xi_{j_l}||)$. Let S (resp. S') be the σ-algebra on Ω spanned by the family of f.r.v. ξ (resp. spanned by the family of r.r.v. $||\xi||$). We will first show that $S' \subset S$.

Let $A \in S'$. Note first that the function $g(.)$ defined from $(\Omega', \mathcal{A}')^{\otimes l}$ to $(\mathbb{R}^+, \mathcal{B}_{\mathbb{R}+})^{\otimes l}$ by $g(\xi) = ||\xi||$ is measurable. The measurability of the variable $g(\xi)$ allows us to write that

$$\exists B \in \mathcal{B}_{\mathbb{R}+}^{\otimes l}, \ A = (g \circ \xi)^{-1}(B) = \xi^{-1} \circ g^{-1}(B).$$

On the other hand, the measurability of g allows us to write that, for this B:

$$\exists C \in \mathcal{A}'^{\otimes l}, \ C = g^{-1}(B).$$

In such a way that $A \in S$, since we have proved that:

$$\exists C \in \mathcal{A}'^{\otimes l}, \ A = \xi^{-1}(C).$$

So we have now proved that $S' \subset S$. This result implies that, if $\alpha_\xi(n)$ (resp. $\alpha_{g(\xi)}(n)$) denotes the coefficients of the sequence $(\xi_n)_{n \in \mathbb{Z}}$ (resp. of the sequence $(||\xi_n||)_{n \in \mathbb{Z}})$, we have:

$$\forall n, \alpha_\xi(n) \geq \alpha_{g(\xi)}(n),$$

which is enough to prove all the assertions of this proposition. $\square$

Proposition 10.4. *Assume that Ω' is a semi-metric space with semi-metric d, and that $\mathcal{A}'$ is the σ-algebra spanned by the open balls for this semi-metric. Let x be a fixed element of Ω' and put $X_i = d(\xi_i, x)$. Then we have:*

i) $(\xi_n)_{n \in \mathbb{Z}}$ is α-mixing $\Rightarrow$ $(d(\xi_n, x))_{n \in \mathbb{Z}}$ is α-mixing;
ii) In addition, if the coefficients of $(\xi_n)_{n \in \mathbb{Z}}$ are geometric (resp. arithmetic) then those of $(d(\xi_n, x))_{n \in \mathbb{Z}}$ are also geometric (resp. arithmetic with the same order).

Proof. Use the same notation as in the proof of Proposition 10.3, with the only change that g is now defined by $g(\xi) = (d(\xi_{j_1}, x), \dots d(\xi_{j_l}, x))$. This being again a measurable function, all the arguments developed before in the proof of Proposition 10.3 remain valid. $\square$

10.4 Mixing and Nonparametric Functional Statistics

The aim of Part IV of our book is to show how the mixing ideas can be combined with the nonparametric kernel functional statistical methods presented in Parts II and III. Theoretical advances will be developed in Chapter 11 while computational issues will be discussed in Chapter 12. Before doing that, let us recall or fix some notations and give a specific formulation of Proposition 10.4 that will be useful throughout the remainder of our book.

As before in this book, we will consider a statistical functional sample $\boldsymbol{\mathcal{X}}_1, \ldots, \boldsymbol{\mathcal{X}}_n$ composed of functional variables each having the same distribution as a generic functional variable $\boldsymbol{\mathcal{X}}$. This functional variable $\boldsymbol{\mathcal{X}}$ takes values in some semi-metric space (E, d). In addition χ (resp. y) is a fixed element of E (resp. of $\mathbb{R}$), $\mathcal{N}_\chi$ $(\subset E)$ is a neighbourhood of χ and S is a fixed compact subset of $\mathbb{R}$. The main novelty in the rest of Part IV is that the data $\boldsymbol{\mathcal{X}}_1, \ldots, \boldsymbol{\mathcal{X}}_n$ will not be assumed to be independent but to be n consecutive terms of a sequence satisfying the strong mixing condition presented in Definition 10.1. In all the following, we will denote by $\alpha(n)$ the mixing coefficients associated with this functional mixing sequence.

As before, K being some kernel function, h being some real positive smoothing parameter, one will have to consider the following locally weighted variables (see general discussion in Chapter 4):

$$\Delta_i = \frac{K\left(\frac{d(\chi, \boldsymbol{\mathcal{X}}_i)}{h}\right)}{\mathbb{E}\left(K\left(\frac{d(\chi, \boldsymbol{\mathcal{X}}_i)}{h}\right)\right)}. \tag{10.1}$$

By applying directly Proposition 10.4, we get the following result that will be used throughout the following chapters.

Lemma 10.5.

i) $(\Delta_i)_{i=1,\ldots n}$ *are n consecutive terms of some α-mixing sequence of r.r.v.;*

ii) *In addition, if the coefficients of the f.r.v. $(\boldsymbol{\mathcal{X}}_i)_{i=1,\ldots n}$ are geometric (resp. arithmetic) then those of $(\Delta_i)_{i=1,\ldots n}$ are also geometric (resp. arithmetic with the same order).*

11

Some Selected Asymptotics

The aim of this chapter is to give extensions to dependent data of several asymptotic results presented in Parts II and III of this book. This chapter is exclusively theoretical, while statistical applications in time series analysis and computational issues are reported in Chapter 12. Sections 11.2, 11.3 and 11.4 will concern the question of predicting some real-valued random response given a functional explanatory variable. Along these previous sections, some auxiliary results about conditional distribution estimation are stated. Section 11.5 is specially devoted to the presentation of these results. Then, Section 11.6 will be concerned with the discrimination problem. Our main goal in this chapter is to show how the dependence is acting on the asymptotic behaviour of the nonparametric functional methods. So, we have decided to present the results by emphasizing (on the hypothesis, as well as on the statement or on the proofs of the results) what is new with α-mixing variables compared with the standard i.i.d. case.

11.1 Introduction

There are always great motivations for studying the behaviour of any statistical method when the usual independence condition on the statistical sample is relaxed. The main reason for this comes from the wish to consider statistical problems involving time series. Of course, this question also occurs with nonparametric functional methods. The aim of this chapter is to provide some theoretical supports about the behaviour on dependent samples of the methods proposed in previous parts of this book. This will be done by means of some almost complete convergence results under mixing dependence modelling that will show the good theoretical behavior of the kernel methods for functional dependent statistical samples. A similar idea, but from a practical point of view, will be supported in Chapter 12.

The chapter is organized as follows: Sections 11.2, 11.3 and 11.4 will cover the question of predicting some real valued random response given a functional explanatory variable. Each of these three sections will attack this problem by means respectively of regression estimation, of functional conditional mode estimation and of functional conditional quantile estimation. Nonparametric estimation is carried out by means of kernel methods, and complete convergence type results will be stated under some strong mixing assumption on the statistical sample. Section 11.5 will present asymptotic results for kernel estimation of conditional density and c.d.f. under mixing assumption. In other words, these four sections will show how the results stated in Chapter 6 for i.i.d. variables remain true in dependent situations. In the same spirit, Section 11.6 is concerned with the discrimination problem. By means of almost complete type results, it will be shown how the kernel supervised classification method behaves asymptotically for discriminating a sample of mixing curves. In other words, Section 11.6 will extend to mixing samples the results stated in Chapter 8 for i.i.d. variables. As for the independent case, the convergence properties are obtained under continuity-type models whereas rates of convergence need Lipschitz-type models.

Our main goal in this chapter is to show how the dependence is acting on the asymptotic behaviour of the nonparametric functional methods. So, we have decided to present our results by emphasizing what is new with mixing variables compared with the standard i.i.d. case. This concerns the presentation of the hypothesis and the statements of the main results. We will highlight how the mixing coefficients are changing (or not) the rates of convergence of the estimates. In the same spirit, our proofs will be rather short (but complete), since we only have to pay attention to the parts for which the dependence structure has some effect (basically the covariance terms in our asymptotic expansions). The other parts (basically bias and variance terms) are behaving as for i.i.d. variables and they will be quickly treated just by refering to previous parts of this book. Of course, such a synthetic presentation has the drawback of obliging the reader to keep in mind previous chapters. However, it has the great advantage of highlighting the influence of the dependence structure on the rates of convergence. It also helps avoid useless repetitions of tedious calculous and notations.

11.2 Prediction with Kernel Regression Estimator

11.2.1 Introduction and Notation

We wish to attack the same problem as described in Chapter 5 but under some dependence assumption on the statistical variables. Precisely, we have to predict a scalar response Y from a functional predictor $\mathcal{X}$ and we will use the nonlinear regression operator r defined by:

$$r(\chi) = \mathbb{E}(Y|\mathcal{X} = \chi). \tag{11.1}$$

Recall that $\mathcal{X}$ is a functional random variable valued in some semi-metric space (E, d), Y is a real random variable, and χ is a fixed element of E.

Let $(\mathcal{X}_i, Y_i)_{i=1,\ldots,n}$ be n pairs being identically distributed as $(\mathcal{X}, Y)$ and satisfying the strong mixing condition introduced in Definition 10.1.

As motivated in Section 5.4 for independent variables, the following kernel estimator of the nonlinear operator r can be proposed:

$$\widehat{r}(\chi) = \frac{\sum_{i=1}^{n} Y_i\, K\left(h^{-1}\, d(\chi, \mathcal{X}_i)\right)}{\sum_{i=1}^{n} K\left(h^{-1}\, d(\chi, \mathcal{X}_i)\right)}, \tag{11.2}$$

where K is an asymmetrical kernel and h (depending on n) is a strictly positive real.

11.2.2 Complete Convergence Properties

The first result of this section is Theorem 11.1 below which is stated in a quite general fashion. This result extends both Theorems 6.1 and 6.11 to mixing variables. The only change, in comparison with what hapens in i.i.d. setting, is appearing through a modification of the second part in the rates of convergence. This comes from the fact that under dependence modelling this second term is now including covariance terms, while the first term (which corresponds to bias effects) is not stochastic and therefore is not affected by the suppression of the independence assumption between the variables. Our Theorem 11.1 is presented in a general way in which the influence of the mixing structure is controled by the following quantities:

$$s_{n,2} = \sum_{i=1}^{n} \sum_{j=1}^{n} cov(Y_i \Delta_i, Y_j \Delta_j), \tag{11.3}$$

$$s_{n,1} = \sum_{i=1}^{n} \sum_{j=1}^{n} cov(\Delta_i, \Delta_j), \tag{11.4}$$

where

$$\Delta_i = \frac{K\left(h^{-1} d(\chi, \mathcal{X}_i)\right)}{\mathbb{E}\, K\left(h^{-1} d(\chi, \mathcal{X}_1)\right)}.$$

In the following, s_n will denote some sequence of positive integers. This sequence will differ according to each result presented below and it will be explicitly specified in each situation, but to fix the ideas let us just say here it will be one among the $s_{n,j}$ defined just before. In comparison with the i.i.d. results given in Theorems 6.1 and 6.11 some additional assumptions are needed in order to control the covariance effects. They consist in assuming either that

$$\left\{ \begin{array}{c} (\mathcal{X}_i, Y_i)_{i=1,\ldots,n} \text{ are strongly mixing} \\ \text{with arithmetic coefficients of order } a > 1, \text{ and} \\ \exists \theta > 2, s_n^{-(a+1)} = o(n^{-\theta}), \end{array} \right. \tag{11.5}$$

or that

$$\begin{cases} (\boldsymbol{X}_i, Y_i)_{i=1,\dots,n} \text{ are strongly mixing} \\ \quad \text{with geometric coefficients, and} \\ \quad \exists \theta > 1, s_n^{-1} = o(n^{-\theta}). \end{cases} \quad (11.6)$$

From a technical point of view it is worth noting that, because of Lemma 10.5, the variables Δ_i (as well as the variables $Y_i \Delta_i$) are mixing real random variables. This will be used implicitly throughout the remainder of the chapter. In particular, this fact makes it possible to apply any among the probabilistic inequalities for real variables described in Section A.3 of Appendix A.

Theorem 11.1. *Put $s_n = \max\{s_{n,2}, s_{n,1}\}$ and assume that either (11.5) or (11.6) is satisfied, then:*
i) Under the conditions of Theorem 6.1 we have:

$$\lim_{n \to \infty} \widehat{r}(\chi) = r(\chi), \quad a.co.,$$

ii) Under the conditions of Theorem 6.11 we have:

$$\widehat{r}(\chi) - r(\chi) = O\left(h^\beta\right) + O_{a.co.}\left(\frac{\sqrt{s_n^2 \log n}}{n}\right).$$

Proof. Let us use the notation introduced in the decomposition (6.7). Because we can write

$$\widehat{r}_2(\chi) = \frac{1}{n}\sum_{i=1}^{n} Y_i \Delta_i,$$

the condition (11.5) (resp. the condition (11.6)) allows to apply a Fuk-Nagaev exponential inequality . More precisely, Corollary A.12-i (resp. Corollary A.13-i) leads directly to:

$$\widehat{r}_2(\chi) - \mathbb{E}\widehat{r}_2(\chi) = O_{a.co.}\left(\frac{\sqrt{s_{n,2}^2 \log n}}{n}\right). \quad (11.7)$$

By the same arguments, since

$$\widehat{r}_1(\chi) = \frac{1}{n}\sum_{i=1}^{n} \Delta_i, \quad (11.8)$$

we have

$$\mathbb{E}\widehat{r}_1(\chi) = 1,$$

and

$$\widehat{r}_1(\chi) - 1 = O_{a.co.}\left(\frac{\sqrt{s_{n,1}^2 \log n}}{n}\right). \quad (11.9)$$

Finally the proof of Theorem 11.1-i (resp. of Theorem 11.1-ii) follows directly from (6.7), (11.7), (11.9) and from Lemma 6.2 (resp. from Lemma 6.12). $\square$

The general results given in Theorem 11.1 before can be formulated in several different specific ways, according to the information we have about the covariance term s_n^2. This term depends directly on the properties of the joint distribution of two distinct pairs $(\boldsymbol{X}_i, Y_i)$ and $(\boldsymbol{X}_j, Y_j)$ and on the behaviour of the mixing coefficients. To fix the ideas and to stay with simple formulations of the results, we will just give below two specific sets of conditions (one for geometric and one for arithmetic coefficients) under which this covariance term has the same behavior as in i.i.d. setting. These results on the covariance term s_n^2 are stated in Lemmas 11.3 and 11.5 below, and their application to the rates of convergence of the nonparametric functional kernel regression estimate are stated in Corollaries 11.4 and 11.6. Other formulations of these corollaries, more general but more complicated in their writing, can be obviously obtained but they are not presented here in order not to mask our main purpose. Some of them can be found in [FV04].

11.2.3 An Application to the Geometrically Mixing Case

We need the following additional assumptions on the distribution of two distinct pairs $(\boldsymbol{X}_i, Y_i)$ and $(\boldsymbol{X}_j, Y_j)$ and on the bandwidth h. We will assume that

$$\forall i \neq j, \mathbb{E}\left(Y_i Y_j | (X_i, X_j)\right) \leq C < \infty, \tag{11.10}$$

and that the joint distribution functions

$$\psi_{i,j}(h) = P\left((\boldsymbol{X}_i, \boldsymbol{X}_j) \in B(\chi, h) \times B(\chi, h)\right),$$

satisfy

$$\exists \epsilon_1 \in (0,1], \, 0 < \psi_\chi(h) = O\left(\varphi_\chi(h)^{1+\epsilon_1}\right), \tag{11.11}$$

where

$$\psi_\chi(h) = \max_{i \neq j} \psi_{i,j}(h).$$

Before going ahead it is worth spending a short moment to discuss the condition (11.11) which may appear quite restrictive but which is indeed much more general than what is usually introduced in finite dimensional setting. This is precisely explicited in the next remark, which is just a special case of Lemma 13.13 that will be stated later on this book.

Remark 11.2. In the special case when the space $E = \mathbb{R}^k$, the condition (11.11) is satisfied with $\epsilon_1 = 1$ as soon as each pair $(\boldsymbol{X}_i, \boldsymbol{X}_j)$ as a density $f_{i,j}$ with respect to the Lebesgues measure on $\mathbb{R}^{2k}$ such that $\sup_{i,j} f_{i,j}(\chi, \chi) \leq C < \infty$.

We also need the following additional condition on the bandwidth:

$$\exists \epsilon_2 \in (0,1), \; \varphi_\chi(h) = O\left(n^{-\epsilon_2}\right). \qquad (11.12)$$

Lemma 11.3. *Assume that the conditions of Theorem 6.11 hold. Assume also that $(\boldsymbol{X}_i, Y_i)_{i=1,\dots,n}$ is a geometrically mixing sequence satisfying (11.10), (11.11) and (11.12). We have for $l = 1$ or 2:*

$$s_{n,l}^2 = O\left(\frac{n}{\varphi_\chi(h)}\right).$$

Proof. • In a first attempt, note that the conditions (11.10) and (11.11) allow to write for any $i \neq j$:

$$\mathbb{E}(Y_i \Delta_i Y_j \Delta_j) = \mathbb{E}\left(\Delta_i \Delta_j \mathbb{E}(Y_i Y_j | (\boldsymbol{X}_i, \boldsymbol{X}_j))\right) \leq C\mathbb{E}\Delta_i \Delta_j$$
$$\leq \frac{C}{\varphi_\chi(h)^2}\mathbb{E}K\left(h^{-1} d(\chi, \boldsymbol{X}_i)\right) K\left(h^{-1} d(\chi, \boldsymbol{X}_j)\right),$$

the last inequality coming from the definition of the Δ_i's and from the result (6.10) applied with $m = 1$. Because K is bounded with compact support, we have directly from the definition of the function $\psi_\chi(.)$:

$$\mathbb{E}(Y_i \Delta_i Y_j \Delta_j) \leq \frac{C}{\varphi_\chi(h)^2} P\left((\boldsymbol{X}_i, \boldsymbol{X}_j) \in B(\chi, h) \times B(\chi, h)\right)$$
$$\leq \frac{C\psi_\chi(h)}{\varphi_\chi(h)^2}.$$

By using now the result (6.9) with $m = 1$ and by introducing the notation,

$$\Psi(h) = \max\{\frac{\psi_\chi(h)}{\varphi_\chi(h)^2}, 1\},$$

we arrive at:

$$|cov(Y_i \Delta_i, Y_j \Delta_j)| \leq |\mathbb{E}(Y_i \Delta_i Y_j \Delta_j)| + |\mathbb{E}(Y_i \Delta_i)||\mathbb{E}(Y_j \Delta_j)|$$
$$\leq O\left(\Psi(h)\right). \qquad (11.13)$$

• Let us treat now the term $s_{n,2}^2$, and write the following decomposition:

$$s_{n,2}^2 = \sum_{i=1}^{n} var(Y_i \Delta_i) + \sum_{0 < |i-j| \leq v_n} \sum cov(Y_i \Delta_i, Y_j \Delta_j)$$
$$+ \sum_{|i-j| > v_n} \sum cov(Y_i \Delta_i, Y_j \Delta_j), \qquad (11.14)$$

where v_n can be any sequence of positive real numbers. The first term in right-hand side of (11.14) can be treated by means of (6.8). The second one can be treated by means of (11.13). The third one can be treated by using the covariance inequality given in Proposition A.10-ii. We arrive finally at:

$$s_{n,2}^2 = O\left(\frac{n}{\varphi_\chi(h)}\right) + O\left(nv_n\Psi(h)\right)$$
$$+ O\left((|\mathbb{E}(Y_i^p\Delta_i^p)||\mathbb{E}(Y_j^p\Delta_j^p)|)^{\frac{1}{p}} \sum\sum_{|i-j|>v_n} \alpha(|j-i|)^{1-\frac{2}{p}}\right). \quad (11.15)$$

By using (6.9) now together with the geometric assumption on the mixing coefficients, we have for some $0 < t < 1$:

$$s_{n,2}^2 = O\left(\frac{n}{\varphi_\chi(h)}\right) + O\left(nv_n\Psi(h)\right)$$
$$+ O\left(\frac{1}{\varphi_\chi(h)^{\frac{2p-2}{p}}} n^2 t^{v_n(1-\frac{2}{p})}\right). \quad (11.16)$$

Note that, because the kernel K is bounded and because Y satisfies the moment condition (6.4), the last result is available for any $p > 2$. Choosing now $v_n = \varphi_\chi(h)^{-\epsilon_1}$ allows us to treat the second term in the right-hand side of (11.16) and to get:

$$s_{n,2}^2 = O\left(\frac{n}{\varphi_\chi(h)}\right) + O\left(\frac{1}{\varphi_\chi(h)^{\frac{2p-2}{p}}} n^2 t^{\frac{p-2}{p\varphi_\chi(h)^{\epsilon_1}}}\right)$$
$$= O\left(\frac{n}{\varphi_\chi(h)}\right) + O\left(\frac{n}{\varphi_\chi(h)}\left(\frac{1}{\varphi_\chi(h)^{\frac{p-2}{p}}}\right) ne^{-\frac{b}{\varphi_\chi(h)^{\epsilon_1}}}\right),$$

for some $b > 0$. So, for any $\eta > 0$, we have:

$$s_{n,2}^2 = O\left(\frac{n}{\varphi_\chi(h)}\right) + O\left(\frac{n}{\varphi_\chi(h)}\left(n\varphi_\chi(h)^\eta\right)\right)$$
$$= O\left(\frac{n}{\varphi_\chi(h)}\right). \quad (11.17)$$

The last result coming directly from the condition (11.12) and by taking η large enough.

• By applying this result to the special case when $Y = 1$, we get

$$s_{n,1}^2 = O\left(\frac{n}{\varphi_\chi(h)}\right). \quad (11.18)$$

So, Lemma 11.3 is proved. □

Corollary 11.4. *Assume that the sequence* $(\mathcal{X}_i, Y_i)_{i=1,\ldots,n}$ *is geometrically mixing. Assume that the conditions of Theorem 6.11 hold together with (11.10), (11.11) and (11.12). Then we have:*

$$\widehat{r}(\chi) - r(\chi) = O\left(h^{\beta}\right) + O_{a.co.}\left(\sqrt{\frac{\log n}{n\varphi_{\chi}(h)}}\right). \tag{11.19}$$

Proof. It suffices to combine Theorem 11.1 with Lemma 11.3. To do that it is necessary to check that condition (11.6) holds, but this follows obviously from (11.12) and Lemma 11.3. $\square$

11.2.4 An Application to the Arithmetically Mixing Case

Let us give now some similar result for arithmetic mixing variables.

Lemma 11.5. *Assume that the conditions of Theorem 6.11 hold, and that the sequence* $(\mathcal{X}_i, Y_i)_{i=1,\ldots,n}$ *is arithmetically mixing with rate satisfying*

$$a > \frac{1 + \epsilon_2}{\epsilon_1 \epsilon_2}. \tag{11.20}$$

If in addition the conditions (11.10), (11.11) and (11.12) are satisfied, we have for $l = 1$ or 2:

$$s_{n,l}^2 = O\left(\frac{n}{\varphi_{\chi}(h)}\right).$$

Proof. We can follow the same steps as for proving Lemma 11.3 before, and so this proof will be presented in a shorter way. Note that the result (11.15) is still valid. So, by using (6.9) together with the arithmetic condition on the mixing coefficients, we get directly from (11.15) that, for any $p > 2$:

$$s_{n,2}^2 = O\left(\frac{n}{\varphi_{\chi}(h)}\right) + O\left(nv_n\Psi(h)\right)$$

$$+ O\left(\frac{1}{\varphi_{\chi}(h)^{\frac{2p-2}{p}}}n^2 v_n^{-a\left(1-\frac{2}{p}\right)}\right)$$

Choosing again $v_n = \varphi_{\chi}(h)^{-\epsilon_1}$ allows us to treat the second term in the right-hand side of (11.21) and to get:

$$s_{n,2}^2 = O\left(\frac{n}{\varphi_{\chi}(h)}\right) + O\left(\frac{1}{\varphi_{\chi}(h)^{\frac{2p-2}{p}}}n^2\varphi_{\chi}(h)^{a\epsilon_1\left(1-\frac{2}{p}\right)}\right)$$

$$= O\left(\frac{n}{\varphi_{\chi}(h)}\right) + O\left(\frac{n}{\varphi_{\chi}(h)}\left(n\varphi_{\chi}(h)^{\frac{(p-2)(a\epsilon_1-1)}{p}}\right)\right)$$

Because of condition (11.20), it is always possible to choose p such that $\epsilon_2 \frac{(p-2)(a\epsilon_1 - 1)}{p} > 1$. So, by choosing such a value for p and by using the condition (11.12), we arrive at:

$$s_{n,2}^2 = O\left(\frac{n}{\varphi_\chi(h)}\right) + O\left(\frac{n}{\varphi_\chi(h)}\left(n^{1-\epsilon_2 \frac{(p-2)(a\epsilon_1-1)}{p}}\right)\right)$$

$$= O\left(\frac{n}{\varphi_\chi(h)}\right).$$

By applying this result to the special case when $Y = 1$ we get directly

$$s_{n,1}^2 = O\left(\frac{n}{\varphi_\chi(h)}\right).$$

The proof of Lemma 11.5 is finished.□

Corollary 11.6. *Assume that the sequence* $(\mathcal{X}_i, Y_i)_{i=1,\dots,n}$ *is arithmetically mixing with order a. Assume that the conditions of Theorem 6.11 hold together with (11.10), (11.11), (11.12) and (11.20). Then we have:*

$$\widehat{r}(\chi) - r(\chi) = O\left(h^\beta\right) + O_{a.co.}\left(\sqrt{\frac{\log n}{n\varphi_\chi(h)}}\right). \qquad (11.21)$$

Proof. It suffices to combine Theorem 11.1 with Lemma 11.5. To do that it is necessary to check that condition (11.5) holds, but this follows obviously from (11.12) and Lemma 11.5. □

11.3 Prediction with Functional Conditional Quantiles

11.3.1 Introduction and Notation

We will now attack the prediction problem of the scalar response Y given the functional predictor $\mathcal{X}$ by mean of functional conditional quantiles. Our aim is to show how the asymptotic results stated in Chapter 6 about conditional c.d.f. and conditional quantiles can be extended to α-mixing functional samples. Before doing that, let us recall some general notations and definitions (see Section 5.2 for details). The nonlinear conditional c.d.f. operator is defined by:

$$\forall y \in \mathbb{R}, \ F_Y^\mathcal{X}(\chi, y) = P(Y \leq y | \mathcal{X} = \chi), \qquad (11.22)$$

while the conditional median, and more generally the conditional quantile of order $\alpha \in (0,1)$ are respectively defined by

$$m(\chi) = \inf\left\{y \in \mathbb{R}, \ F_Y^{\boldsymbol{\mathcal{X}}}(\chi, y) \geq 1/2\right\}, \qquad (11.23)$$

and

$$t_\alpha(\chi) = \inf\left\{y \in \mathbb{R}, \ F_Y^{\boldsymbol{\mathcal{X}}}(\chi, y) \geq \alpha\right\}. \qquad (11.24)$$

As discussed in Section 5.4, kernel smoothing ideas can be used to estimate nonparametrically these nonlinear operators. Precisely, kernel estimates of $F_Y^{\boldsymbol{\mathcal{X}}}$, $t_\alpha(\chi)$ and $m(\chi)$ are respectively defined by:

$$\widehat{F}_Y^{\boldsymbol{\mathcal{X}}}(\chi, y) \ = \ \frac{\sum_{i=1}^n K\left(h^{-1}d(\chi, \boldsymbol{\mathcal{X}}_i)\right) H\left(g^{-1}\left(y - Y_i\right)\right)}{\sum_{i=1}^n K\left(h^{-1}d(\chi, \boldsymbol{\mathcal{X}}_i)\right)}, \qquad (11.25)$$

$$\widehat{t}_\alpha(\chi) \ = \ \inf\left\{y \in \mathbb{R}, \ \widehat{F}_Y^{\boldsymbol{\mathcal{X}}}(\chi, y) \geq \alpha\right\}, \qquad (11.26)$$

and

$$\widehat{m}(\chi) \ = \ \widehat{t}_{\frac{1}{2}}(\chi). \qquad (11.27)$$

In these definitions K is an asymmetrical kernel, H is some integrated kernel, while h and g are nonnegative smoothing parameters (depending on the sample size n).

11.3.2 Complete Convergence Properties

The presentation of the results will follow the same lines as in Section 11.2.2 for regression estimation. To save space, we will present directly the results for the estimation of any conditional quantile $t_\alpha(\chi)$, without stating explicitly results for the conditional median. Of course, they can be directly obtained by taking $\alpha = 1/2$. A previous general result is stated in Theorem 11.7 below, in order to extend both Theorems 6.8 and 6.18 to mixing variables. For the same reasons as in Section 11.2.2, the influence of the mixing structure on the rates of convergence will be seen through the following quantities:

$$s_{n,3,0} = \sum_{i=1}^n \sum_{j=1}^n cov(\Gamma_i(y)\Delta_i, \Gamma_j(y)\Delta_j), \qquad (11.28)$$

$$s_{n,3,l} = \sum_{i=1}^n \sum_{j=1}^n cov(\Gamma_i^{(l)}(y)\Delta_i, \Gamma_j^{(l)}(y)\Delta_j), \qquad (11.29)$$

and

$$s_{n,1} = \sum_{i=1}^n \sum_{j=1}^n cov(\Delta_i, \Delta_j), \qquad (11.30)$$

where

$$\Delta_i \ = \ \frac{K\left(h^{-1}d(\chi, \boldsymbol{\mathcal{X}}_i)\right)}{\mathbb{E}\,K\left(h^{-1}d(\chi, \boldsymbol{\mathcal{X}}_1)\right)},$$

and

$$\Gamma_i(y) = H\left(g^{-1}(y - Y_i)\right).$$

In the following s_n will denote some sequence of positive integers. This sequence will differ according to each result presented below and it will be explicitly specified in each situation, but to fix the ideas let us just say here it will be one among $\{s_{n,1}, s_{n,3,l}, l = 0 \ldots j\}$. In comparison with the i.i.d. results some additional assumptions are needed in order to control these covariance effects. They consist in assuming either that

$$\begin{cases} (\mathcal{X}_i, Y_i)_{i=1,\ldots,n} \text{ are strongly mixing} \\ \text{with arithmetic coefficients of order } a > 1, \text{ and} \\ \exists \theta > 2, s_n^{-(a+1)} = o(n^{-\theta}), \end{cases} \qquad (11.31)$$

or that

$$\begin{cases} (\mathcal{X}_i, Y_i)_{i=1,\ldots,n} \text{ are strongly mixing} \\ \text{with geometric coefficients, and} \\ \exists \theta > 1, s_n^{-1} = o(n^{-\theta}). \end{cases} \qquad (11.32)$$

Theorem 11.7. *i) Put* $s_n = \max\{s_{n,1}, s_{n,3,0}\}$, *and assume that either (11.31) or (11.32) is satisfied together with the conditions of Theorem 6.8. Then we have:*

$$\lim_{n \to \infty} \widehat{t}_\alpha(\chi) = t_\alpha(\chi), \quad a.co.$$

ii) Put $s_n = \max\{s_{n,1}, s_{n,3,l}, l = 0 \ldots j\}$, *and assume that either (11.31) or (11.32) is satisfied together with the conditions of Theorem 6.18. Then we have:*

$$\widehat{t}_\alpha(\chi) - t_\alpha(\chi) = O\left(\left(h^\beta + g^\beta\right)^{\frac{1}{j}}\right) + O_{a.co.}\left(\left(\frac{s_n^2 \log n}{n^2}\right)^{\frac{1}{2j}}\right).$$

Proof. This proof is based on previous results concerning the estimation of conditional c.d.f. nonlinear operator $F_Y^{\mathcal{X}}$ by the kernel estimate $\widehat{F}_Y^{\mathcal{X}}$ (see Lemmas 11.8, 11.9 and 11.10 below).

i) Apart from Lemma 6.5, the proof of Theorem 6.8 was performing along analytic deterministic arguments and is therefore not affected by the suppression of the independence condition between the variables. In other words, the proof of Theorem 11.7-i follows directly along the same lines as the proof of Theorem 6.8, but using Lemma 11.8 below in place of Lemma 6.5.

ii) Similarly, the proof of Theorem 11.7-ii follows directly along the same lines as the proof of Theorem 6.18, but using Lemmas 11.9 and 11.10 below in place of Lemmas 6.14 and 6.15.□

Lemma 11.8. *Under the conditions of Theorem 11.7-i, we have for any fixed real point y:*

$$\lim_{n \to \infty} \widehat{F}_Y^{\mathcal{X}}(y) = F_Y^{\mathcal{X}}(y), \quad a.co.$$

Proof. This proof is performed over the same steps as the proof of Lemma 6.5, and when necessary, the same notation will be used. Because the kernels K and H are supposed to be bounded, the term $\widehat{r}_3(\chi)$ can be written as a sum of bounded mixing variables

$$\widehat{r}_3(\chi) = \frac{1}{n} \sum_{i=1}^{n} \Gamma_i(y) \Delta_i.$$

So, the condition (11.31) (resp. the condition (11.32)) allows to apply Corollary A.12-ii (resp. Corollary A.13-ii) and to get directly that

$$\widehat{r}_3(\chi) - \mathbb{E}\widehat{r}_3(\chi) = O_{a.co.}\left(\frac{\sqrt{s_{n,3,0}^2 \log n}}{n}\right). \tag{11.33}$$

Finally the proof of Lemma 11.8 follows directly from (6.18), (6.19), (11.9) and (11.33). □

Lemma 11.9. *Under the conditions of Theorem 11.7-ii, we have for any fixed real point* y:

$$F_Y^\chi(y) - \widehat{F}_Y^\chi(y) = O\left(h^\beta + g^\beta\right) + O_{a.co.}\left(\frac{\sqrt{(\max\{s_{n,3,0}, s_{n,1}\})^2 \log n}}{n}\right).$$

Proof. This result follows directly from (6.18), (6.75), (11.9) and (11.33). □

Lemma 11.10. *Let* l *be an integer* $l \in \{1, \ldots, j\}$ *and* y *a fixed real number. Under the conditions of Theorem 11.7-ii, we have:*

$$\lim_{n \to \infty} \widehat{F}_Y^{\chi^{(l)}}(y) = F_Y^{\chi^{(l)}}(y), \quad a.co.$$

If in addition the condition (6.81) is satisfied, then we have

$$F_Y^{\chi^{(l)}}(y) - \widehat{F}_Y^{\chi^{(l)}}(y) = O\left(h^\beta + g^{\beta_0}\right) + O_{a.co.}\left(\frac{\sqrt{(\max\{s_{n,3,l}, s_{n,1}\})^2 \log n}}{ng^{2l-1}}\right).$$

Proof. This proof is similar to the proof of Lemma 6.15, and we will use the same notation as those introduced in the decompositions (6.83) and (6.87). It follows from (6.87), (6.88) and (6.89), and by using the Fuk-Nagaev exponential inequality given in Corollary A.13-ii, that

$$\widehat{r}_3^{(l)}(\chi, y) - \mathbb{E}\widehat{r}_3^{(l)}(\chi, y) = O_{a.co.}\left(\frac{\sqrt{s_{n,3,l}^2 \log n}}{n}\right). \tag{11.34}$$

The first assertion of Lemma 11.10 follows directly from (6.83), (6.85), (11.9) and (11.34). Similarly, the second assertion of Lemma 11.10 follows from (6.83), (6.86), (11.9) and (11.34). □

The general result given in Theorem 11.7-ii can be formulated in several different specific ways, according to the knowledge we have about the covariance terms s_n^2. As before in regression (see discussion in Section 11.2), we will only present below two special cases (those for which the covariance terms have the same behaviour as in i.i.d. setting). More general formulations can be found in [FRV05]. These results on the covariance term s_n^2 are stated in Lemmas 11.11 and 11.13 below, and their application to the rates of convergence of the nonparametric kernel conditional quantile estimate are stated in Corollaries 11.12 and 11.14. The conditions are similar to those appearing in regression (see Remark 11.2). Precisely, we will assume that

$$\exists \epsilon_1 \in (0,1),\ 0 < \psi_\chi(h) = O\left(\varphi_\chi(h)^{1+\epsilon_1}\right), \tag{11.35}$$

$$\exists \epsilon_2 \in (0,1),\ g\varphi_\chi(h) = O\left(n^{-\epsilon_2}\right), \tag{11.36}$$

and that for any $i \neq j$:

The conditional density $f_{i,j}$ of (Y_i, Y_j)
given $(\mathcal{X}_i, \mathcal{X}_j)$ exists and is bounded. $\tag{11.37}$

11.3.3 Application to the Geometrically Mixing Case

We will show in this section that the rates of convergence of the nonparametric kernel conditional quantile for geometrically mixing variables can be the same as for i.i.d. ones.

Lemma 11.11. *Assume that the conditions of Theorem 6.18 hold. Assume also that $(\mathcal{X}_i, Y_i)_{i=1,\dots,n}$ is a geometrically mixing sequence satisfying (11.35), (11.36) and (11.37). We have*

$$s_{n,3,0}^2 = O\left(\frac{n}{\varphi_\chi(h)}\right).$$

and for any $l = 1,\dots j$:

$$s_{n,3,l}^2 = O\left(\frac{n}{g^{(2l-1)}\varphi_\chi(h)}\right).$$

Proof. • This proof is performed over the same lines as the proof of Lemma 11.3 above. Some details will be therefore omitted. It is based on the following decomposition, which is valid for any $l = 0,\dots j$.

$$s_{n,3,l}^2 = \sum_{i=1}^n var(\Gamma_i^{(l)}(y)\Delta_i) + \sum_{0<|i-j|\leq v_n}\sum cov\left(\Gamma_i^{(l)}(y)\Delta_i, \Gamma_j^{(l)}(y)\Delta_j\right)$$

$$+ \sum_{|i-j|>v_n}\sum cov\left(\Gamma_i^{(l)}(y)\Delta_i, \Gamma_j^{(l)}(y)\Delta_j\right), \tag{11.38}$$

where v_n can be any sequence of positive real numbers.

- Let us first prove the result of Lemma 11.11 for $l = 0$. Because H is bounded, we can apply directly the result (11.13) to get:

$$cov\left(\Gamma_i(y)\Delta_i, \Gamma_j(y)\Delta_j\right) = O\left(\Psi(h)\right), \qquad (11.39)$$

where

$$\Psi(h) = \max\{\frac{\psi_\chi(h)}{\varphi_\chi(h)^2}, 1\}.$$

The first term on the right-hand side of (11.38) can be treated by means of (6.85) and (6.89). The second one can be treated by means of (11.39). The third one can be treated by using the covariance inequality given in Proposition A.10-ii. We arrive finally at:

$$s_{n,3,0}^2 = O\left(\frac{n}{\varphi_\chi(h)}\right) + O\left(nv_n\Psi(h)\right)$$

$$+ O\left(\frac{1}{(\mathbb{E}\Gamma_i(y)^p\Delta_i^p)^{\frac{2}{p}}}n^2\alpha(v_n)^{1-\frac{2}{p}}\right)$$

$$= O\left(\frac{n}{\varphi_\chi(h)}\right) + O\left(nv_n\Psi(h)\right)$$

$$+ O\left(\frac{1}{\varphi_\chi(h)^{\frac{2p-2}{p}}}n^2\alpha(v_n)^{1-\frac{2}{p}}\right), \qquad (11.40)$$

the last inequality following directly from the boundedness property of H and for the result (6.9) on higher moments of Δ_i. It suffices now to take $v_n = \varphi_\chi(h)^{-\epsilon_1}$ to treat the second term on the right-hand side of (11.15) and to use the geometric condition on the mixing coefficients to get, for some $0 < t < 1$,:

$$s_{n,3,0}^2 = O\left(\frac{n}{\varphi_\chi(h)}\right) + O\left(\frac{1}{\varphi_\chi(h)^{\frac{2p-2}{p}}}n^2 t^{\frac{p-2}{p\varphi_\chi(h)^{\epsilon_1}}}\right)$$

$$= O\left(\frac{n}{\varphi_\chi(h)}\right) + O\left(\frac{1}{\varphi_\chi(h)^{\frac{2p-2}{p}}}n^2 e^{-\frac{b}{\varphi_\chi(h)^{\epsilon_1}}}\right),$$

for some $b > 0$. This is enough, because of condition (11.12), to show that

$$s_{n,3,0}^2 = O\left(\frac{n}{\varphi_\chi(h)}\right).$$

- Let us now prove the result of Lemma 11.11 for $l \in \{1, \ldots j\}$. Using the same arguments as those invoked to prove (6.89), and using the condition (11.37) on the joint conditional distribution, we have for any $i \neq j$:

$$\mathbb{E}\Gamma_i^{(l)}(y)\Delta_i\Gamma_j^{(l)}(y)\Delta_j =$$

$$= \frac{1}{g^{2l}} \mathbb{E} \left(\Delta_i \Delta_j \int_{\mathbb{R}^2} H^{(l)}(\frac{y-u_1}{g}) H^{(l)}(\frac{y-u_2}{g}) f_{i,j}(u_1, u_2) du_1 du_2 \right)$$

$$= \frac{1}{g^{2l-2}} \mathbb{E} \left(\Delta_i \Delta_j \int_{\mathbb{R}^2} H^{(l)}(v_1) H^{(l)}(v_2) f_{i,j}(y-gv_1, y-gv_2) dv_1 dv_2 \right)$$

$$\le C \frac{1}{g^{2l-2}} \mathbb{E} \Delta_i \Delta_j$$

$$\le C \frac{1}{g^{2l-2}} \frac{1}{\varphi(h)^2} P((\mathcal{X}_i, \mathcal{X}_j) \in B(\chi, h) \times B(\chi, h))$$

$$= O \left(\frac{\psi_{i,j}(h)}{\varphi(h)^2 g^{2l-2}} \right). \tag{11.41}$$

By the same kind of aguments, we have for any $p > 0$:

$$\mathbb{E} \Gamma_i^{(l)}(y)^p \Delta_i^p = \frac{1}{g^{pl}} \mathbb{E} \left(\Delta_i^p \int_{\mathbb{R}} H^{(l)}(\frac{y-u}{g})^p f_Y^{\mathcal{X}}(u) du \right)$$

$$= \frac{1}{g^{pl-1}} \mathbb{E} \left(\Delta_i^p \int_{\mathbb{R}^2} H^{(l)}(v)^p f_Y^{\mathcal{X}}(y-v) dv \right)$$

$$\le C \frac{1}{g^{pl-1}} \mathbb{E} \Delta_i^p$$

$$= O \left(\frac{1}{g^{pl-1}} \frac{1}{\varphi(h)^{p-1}} \right), \tag{11.42}$$

where the last bound comes directly from the inequality (6.11). Finally, we get directly from (11.41) and (11.42):

$$cov(\Gamma_i^{(l)}(y) \Delta_i, \Gamma_j^{(l)}(y) \Delta_j) = O \left(\frac{\Psi(h)}{g^{2l-2}} \right). \tag{11.43}$$

On the other hand, the result (11.42) together with the covariance inequality given in Proposition A.10-ii, leads directly for any $p > 0$ to:

$$cov(\Gamma_i^{(l)}(y) \Delta_i, \Gamma_j^{(l)}(y) \Delta_j) = O \left(\frac{1}{\mathbb{E}(\Gamma_1^{(l)}(y) \Delta_1)^{2p}} \alpha(|i-j|)^{(1-\frac{2}{p})} \right)$$

$$= O \left(\frac{1}{g^{\frac{2(pl-1)}{p}}} \frac{1}{\varphi_\chi(h)^{\frac{2p-2}{p}}} \alpha(|i-j|)^{(1-\frac{2}{p})} \right). \tag{11.44}$$

Recall that the variance term is the same as in the i.i.d. setting, and so we have directly from (6.85) and (6.89) that:

$$var(\Gamma_i^{(l)}(y) \Delta_i) = O \left(\frac{1}{g^{2l-1} \varphi_\chi(h)} \right). \tag{11.45}$$

The first term in the right-hand side of (11.38) can be treated by (11.45). The second one can be treated by means of and (11.43), while the third one can be treated by using (11.44). We arrive finally at:

$$s_{n,3,l}^2 = O\left(\frac{n}{g^{2l-1}\varphi_\chi(h)}\right) + O\left(nv_n\frac{\Psi(h)}{g^{2l-2}}\right)$$

$$+ O\left(\frac{1}{g^{\frac{2(pl-1)}{p}}}\frac{1}{\varphi_\chi(h)^{\frac{2p-2}{p}}}n^2\alpha(v_n)^{1-\frac{2}{p}}\right). \qquad (11.46)$$

By choosing $v_n = g^{-1}\varphi_\chi(h)^{-\epsilon_1}$ and according to (11.35) the second term on the right-hand side of (11.46) becomes of lower order than the first one, and because the mixing coefficients are geometrically decaying we arrive, for some $0 < t < 1$ and for some $b > 0$, at:

$$s_{n,3,l}^2 = O\left(\frac{n}{g^{2l-1}\varphi_\chi(h)}\right) + O\left(\frac{1}{g^{\frac{2(pl-1)}{p}}}\frac{1}{\varphi_\chi(h)^{\frac{2p-2}{p}}}n^2 t^{\frac{p-2}{pg\varphi_\chi(h)^{\epsilon_1}}}\right)$$

$$= O\left(\frac{n}{g^{2l-1}\varphi_\chi(h)}\right) + O\left(\frac{1}{g^{\frac{2(pl-1)}{p}}}\frac{1}{\varphi_\chi(h)^{\frac{2p-2}{p}}}n^2 e^{-\frac{b}{g\varphi_\chi(h)^{\epsilon_1}}}\right)$$

$$= O\left(\frac{n}{g^{2l-1}\varphi_\chi(h)}\right) + O\left(\frac{n}{g^{2l-1}\varphi_\chi(h)}\left(\frac{1}{(g\varphi_\chi(h))^{\frac{p-2}{p}}}ne^{-\frac{b}{g\varphi_\chi(h)^{\epsilon_1}}}\right)\right).$$

So we have for any $\eta > 0$:

$$s_{n,3,l}^2 = O\left(\frac{n}{g^{2l-1}\varphi_\chi(h)}\right) + O\left(\frac{n}{g^{2l-1}\varphi_\chi(h)}\left(\frac{1}{(g\varphi_\chi(h))^{\frac{p-2}{p}}}n(g\varphi_\chi(h))^\eta\right)\right),$$

which is enough, because of condition (11.36), to show that

$$s_{n,3,l}^2 = O\left(\frac{n}{g^{2l-1}\varphi_\chi(h)}\right) + O\left(\frac{n}{g^{2l-1}\varphi_\chi(h)}n^{1-\epsilon_2}\eta^{\frac{p-2}{p}}\right).$$

Taking η large enough, we arrive at:

$$s_{n,3,l}^2 = O\left(\frac{n}{g^{2l-1}\varphi_\chi(h)}\right).$$

$\square$

Corollary 11.12. *Assume that the sequence* $(\mathcal{X}_i, Y_i)_{i=1,\dots,n}$ *is geometrically mixing. Assume that the conditions of Theorem 6.18 hold together with (11.35), (11.36) and (11.37). Then we have:*

$$\widehat{t}_\alpha(\chi) - t_\alpha(\chi) = O\left(\left(h^\beta + g^\beta\right)^{\frac{1}{j}}\right) + O_{a.co.}\left(\left(\frac{\log n}{n\varphi_\chi(h)}\right)^{\frac{1}{2j}}\right).$$

Proof. By combining the results of Lemma 11.9, Lemma 11.10 and Lemma 11.11, we arrive directly at both of the following results which can be seen as extensions to arithmetically mixing variables of the Lemmas 6.14 and 6.15 stated before in the i.i.d. setting:

$$F_Y^X(y) - \widehat{F}_Y^X(y) = O\left(h^\beta + g^\beta\right) + O_{a.co.}\left(\left(\frac{\log n}{n\varphi_\chi(h)}\right)^{\frac{1}{2}}\right),$$

and

$$F_Y^{X^{(l)}}(y) - \widehat{F}_Y^{X^{(l)}}(y) = O\left(h^\beta + g^{\beta_0}\right) + O_{a.co.}\left(\left(\frac{\log n}{ng^{2l-1}\varphi_\chi(h)}\right)^{\frac{1}{2}}\right).$$

So, the proof of Corollary 11.12 follows exactly along the same lines as the proof of Theorem 6.8, but using these two previous results in place of Lemmas 6.14 and 6.15.□

11.3.4 Application to the Arithmetically Mixing Case

We will give now similar results but for variables satisfying the arithmetic mixing condition.

Lemma 11.13. *Assume that the conditions of Theorem 6.18 hold. Assume also that (11.35), (11.36) and (11.37) are satisfied and that $(\mathcal{X}_i, Y_i)_{i=1,\dots,n}$ is an arithmetically mixing sequence with order such that:*

$$a > \frac{1 + \epsilon_1}{\epsilon_1 \epsilon_2}. \tag{11.47}$$

We have

$$s_{n,3,0}^2 = O\left(\frac{n}{\varphi_\chi(h)}\right),$$

and for any $l = 1, \dots j$:

$$s_{n,3,l}^2 = O\left(\frac{1}{g^{(2l-1)}}\frac{n}{\varphi_\chi(h)}\right).$$

Proof. This proof is performed over the same steps as for proving Lemma 11.11 before, and is based on the decomposition (11.38).

• Let us first prove the result of Lemma 11.13 for $l = 0$. Note that the result (11.40) is still valid in the arithmetic mixing setting, since its statement did not need any specific form for the mixing coefficients. It suffices to take $v_n = \varphi_\chi(h)^{-\epsilon_1}$ to treat the second term on the right-hand side of (11.15) and to use the arithmetic condition on the mixing coefficients, to get:

$$s_{n,3,0}^2 = O\left(\frac{n}{\varphi_\chi(h)}\right) + O\left(\frac{1}{\varphi_\chi(h)^{\frac{2p-2}{p}}} n^2 \varphi_\chi(h)^{a\epsilon_1\left(1-\frac{2}{p}\right)}\right)$$

$$= O\left(\frac{n}{\varphi_\chi(h)}\right) + O\left(\frac{n}{\varphi_\chi(h)}\left(n\varphi_\chi(h)^{\frac{(p-2)(a\epsilon_1-1)}{p}}\right)\right)$$

$$= O\left(\frac{n}{\varphi_\chi(h)}\right) + O\left(\frac{n}{\varphi_\chi(h)}\left(n^{1-\epsilon_2\frac{(p-2)(a\epsilon_1-1)}{p}}\right)\right).$$

This result being true for any p and because of condition (11.47), it is always possible to choose p such that $\epsilon_2 \frac{(p-2)(a\epsilon_1-1)}{p} > 1$, and so we have the claimed result

$$s_{n,3,0}^2 = O\left(\frac{n}{\varphi_\chi(h)}\right).$$

• Let us now prove the result of Lemma 11.13 for $l \in \{1, \ldots j\}$. Note that the result (11.46) is still valid. By choosing $v_n = g^{-1}\varphi_\chi(h)^{-\epsilon_1}$ we can treat the second term on the right-hand side of (11.46), and because the mixing coefficients are arithmetically decreasing with order a, we arrive at:

$$s_{n,3,l}^2 = O\left(\frac{n}{g^{2l-1}\varphi_\chi(h)}\right)$$

$$+ O\left(\frac{1}{g^{\frac{2(pl-1)}{p}}} \frac{1}{\varphi_\chi(h)^{\frac{2p-2}{p}}} n^2 (g\varphi_\chi(h)^{\epsilon_1})^{a\left(1-\frac{2}{p}\right)}\right)$$

$$= O\left(\frac{n}{g^{2l-1}\varphi_\chi(h)}\right)$$

$$+ O\left(\frac{n}{g^{2l-1}\varphi_\chi(h)}\left(\frac{1}{(g\varphi_\chi(h))^{\frac{p-2}{p}}} n(g\varphi_\chi(h))^{a\epsilon_1\left(\frac{p-2}{p}\right)}\right)\right)$$

This result is true for any p. In the other hand, because of condition (11.47), it is always possible to choose p such that

$$\epsilon_2 \frac{(p-2)(a\epsilon_1-1)}{p} > 1.$$

So, by choosing such a value for p and by using condition (11.36), we have the claimed result

$$s_{n,3,l}^2 = O\left(\frac{n}{g^{2l-1}\varphi_\chi(h)}\right) + O\left(\frac{n}{g^{2l-1}\varphi_\chi(h)} n^{1-\epsilon_2\left(\frac{(p-2)(a\epsilon_1-1)}{p}\right)}\right)$$

$$= O\left(\frac{n}{g^{2l-1}\varphi_\chi(h)}\right).$$

$\square$

Corollary 11.14. *Assume that the sequence* $(\mathcal{X}_i, Y_i)_{i=1,\ldots,n}$ *is arithmetically mixing with rates satisfying (11.47). Assume that the conditions of Theorem 6.18 hold together with (11.35), (11.36) and (11.37). Then we have:*

$$\widehat{t_\alpha}(\chi) - t_\alpha(\chi) = O\left(\left(h^\beta + g^\beta\right)^{\frac{1}{j}}\right) + O_{a.co.}\left(\left(\frac{\log n}{n\varphi_\chi(h)}\right)^{\frac{1}{2j}}\right).$$

Proof. By combining the results of Lemma 11.9, Lemma 11.10 and Lemma 11.13, we arrive directly at both of the following results which can be seen as extensions to arithmetically mixing variables of the Lemmas 6.14 and 6.15 stated before in the i.i.d. setting:

$$F_Y^\chi(y) - \widehat{F}_Y^\chi(y) = O\left(h^\beta + g^\beta\right) + O_{a.co.}\left(\left(\frac{\log n}{n\varphi_\chi(h)}\right)^{\frac{1}{2}}\right),$$

and

$$F_Y^{\chi(l)}(y) - \widehat{F}_Y^{\chi(l)}(y) = O\left(h^\beta + g^{\beta_0}\right) + O_{a.co.}\left(\left(\frac{\log n}{ng^{2l-1}\varphi_\chi(h)}\right)^{\frac{1}{2}}\right).$$

So, the proof of Corollary 11.14 follows exactly along the same lines as the proof of Theorem 6.8, but using these two previous results in place of Lemmas 6.14 and 6.15.$\square$

11.4 Prediction with Conditional Mode

11.4.1 Introduction and Notation

We will now attack the prediction problem of the scalar response Y given the functional predictor $\mathcal{X}$ by means of functional conditional mode, and we will see how the asymptotic results stated in Chapter 6 about conditional density and conditional modes behave for α-mixing functional samples. Let us recall some general notations and definitions (see Section 5.2 for details). The nonlinear conditional density operator is defined to be the density of the conditional c.d.f. $F_Y^{\mathcal{X}}$ defined in (11.22). Under differentiability assumption, it can be written as:

$$\forall y \in \mathbb{R}, \ f_Y^{\mathcal{X}}(\chi, y) = \frac{\partial}{\partial y} F_Y^{\mathcal{X}}(\chi, y). \tag{11.48}$$

The conditional mode, which is assumed to exist on some subset $S \subset \mathbb{R}$, is defined by:

$$\theta(\chi) = \arg\sup_{y \in S} f_Y^{\mathcal{X}}(\chi, y). \tag{11.49}$$

As discussed in Section 5.4, kernel smoothing ideas can be used to estimate these operators nonparametrically. Kernel estimates of $f_Y^{\mathcal{X}}$ and $\theta(\chi)$ are respectively defined by:

$$\widehat{f_Y^{\mathcal{X}}}(\chi, y) = \frac{g^{-1} \sum_{i=1}^{n} K\left(h^{-1} d(\chi, \mathcal{X}_i)\right) K_0\left(g^{-1}(y - Y_i)\right)}{\sum_{i=1}^{n} K\left(h^{-1} d(\chi, \mathcal{X}_i)\right)}, \tag{11.50}$$

and

$$\widehat{\theta}(\chi) = \arg\sup_{y \in S} \widehat{f_Y^{\mathcal{X}}}(\chi, y). \tag{11.51}$$

In these defnitions K is an asymmetrical kernel, K_0 is a symmetrical kernel, while h and g are nonnegative smoothing parameters (depending on the sample size n).

11.4.2 Complete Convergence Properties

The presentation of the results will follow the same lines as in Section 11.2.2 for regression and in Section 11.3.2 for quantile estimation. A previous general result is stated in Theorem 11.15 below, in order to extend both Theorems 6.6 and 6.16 to mixing variables. As before, the influence of the mixing structure on the rates of convergence will be seen through the following quantities:

$$s_{n,4} = \sum_{i=1}^{n} \sum_{j=1}^{n} cov(\Omega_i(y)\Delta_i, \Omega_j(y)\Delta_j), \tag{11.52}$$

and

$$s_{n,1} = \sum_{i=1}^{n} \sum_{j=1}^{n} cov(\Delta_i, \Delta_j), \tag{11.53}$$

where

$$\Delta_i = \frac{K\left(h^{-1}d(\chi, \mathcal{X}_i)\right)}{\mathbb{E}\, K\left(h^{-1}d(\chi, \mathcal{X}_1)\right)},$$

and

$$\Omega_i(y) = g^{-1}K_0\left(g^{-1}(y - Y_i)\right).$$

In comparison with the i.i.d. results we need some additional assumptions. In the following, s_n denotes some sequence of positive integers. This sequence will differ according to each result presented below and it will be explicitly specified in each situation, but to fix the ideas let us just say here that it will be one among the $s_{n,j}$ defined just before. Recall that ζ was defined along the condition (6.31). Firstly, we will assume either that

$$\begin{cases} (\mathcal{X}_i, Y_i)_{i=1,\dots,n} \text{ are strongly mixing} \\ \text{with arithmetic coefficients of order } a > 1, \text{ and} \\ \exists \theta > 2(1 + \zeta), s_n^{-(a+1)} = o(n^{-\theta}), \end{cases} \tag{11.54}$$

or that

$$\begin{cases} (\mathcal{X}_i, Y_i)_{i=1,\dots,n} \text{ are strongly mixing} \\ \quad \text{with geometric coefficients, and} \\ \quad \exists \theta > 1, s_n^{-1} = o(n^{-\theta}). \end{cases} \qquad (11.55)$$

Theorem 11.15. *i) Put* $s_n = \max\{s_{n,1}, s_{n,4}\}$, *and assume that either* *(11.54) or (11.55) is satisfied together with the conditions of Theorem 6.6.* *Then we have:*

$$\lim_{n \to \infty} \widehat{\theta}(\chi) = \theta(\chi), \quad a.co.$$

ii) Put $s_n = \max\{s_{n,1}, s_{n,3,l}, l = 0 \dots j\}$, *and assume that either (11.54) or* *(11.55) is satisfied together with the conditions of Theorem 6.16. Then we* *have:*

$$\widehat{\theta}(\chi) - \theta(\chi) = O\left(\left(h^\beta + g^\beta \right)^{\frac{1}{j}} \right) + O_{a.co.}\left(\left(\frac{s_n^2 \log n}{n^2 g} \right)^{\frac{1}{2j}} \right).$$

Proof. This proof is based on previous results concerning the kernel estimate $\widehat{f}_Y^\mathcal{X}$ of the nonlinear conditional density operator (see Lemmas 11.16 and 11.17 below).
i) Apart from Lemma 6.7, the proof of Theorem 6.6 was performed only along analytic deterministic arguments and is therefore not affected by the suppression of the independence condition between the variables. So, the only thing to do is to prove that the result of Lemma 6.7 remains true in mixing situations. This will be done in Lemma 11.16 below. In other words, the proof of Theorem 11.15-i follows directly from (6.33) and Lemma 11.16.
ii) Similarly, the proof of Theorem 11.15-ii needs an extension of Lemma 6.17 to dependent variables. This will be done in Lemma 11.17 below. Said differently, the result of Theorem 11.15-ii follows directly from (6.98) and by applying Lemma 11.17 below with $l = 1$ and $S = (\theta(\chi) - \xi, \theta(\chi) + \xi)$. Note that this is valid since $s_{n,4} = s_{n,3,1}$. $\square$

Lemma 11.16. *Under the conditions of Theorem 11.15-i, we have for any* *compact subset* $S \subset \mathbb{R}$:

$$\lim_{n \to \infty} \sup_{y \in S} |f_Y^\mathcal{X}(y) - \widehat{f}_Y^\mathcal{X}(y)| = 0, \quad a.co.$$

Proof. • This proof follows the same lines as the proof of Lemma 6.7, and we wil present it in a rather brief way. Indeed, we will emphasize those among the steps of the proof which are affected by the new dependence situation. Let us use the same notation as in Lemma 6.7. Because the decomposition (6.35) remains valid, our proof will be complete as long as both of the following properties can be checked:

$$\lim_{n\to\infty} \frac{1}{\widehat{r}_1(\chi)} \sup_{y\in S} |\mathbb{E}\widehat{r}_4(\chi,y) - f_Y^X(y)| = 0, \ a.co., \tag{11.56}$$

and

$$\lim_{n\to\infty} \frac{1}{\widehat{r}_1(\chi)} \sup_{y\in S} |\widehat{r}_4(\chi,y) - \mathbb{E}\widehat{r}_4(\chi,y)| = 0, \ a.co. \tag{11.57}$$

- Because the bias terms are not affected with the new dependent situation, the result (6.40) remains true. So, the result (11.56) is obtained directly by combining (6.40), (11.9) and Proposition A.5-i.
- To check (11.57), we can write that $S \subset \bigcup_{k=1}^{z_n} S_k$, where $S_k = (t_k - l_n, t_k + l_n)$ and where $l_n = Cz_n^{-1} = n^{-2a}$. We will use the decomposition

$$\frac{1}{\widehat{r}_1(\chi)} \sup_{y\in S} |\widehat{r}_4(\chi,y) - \mathbb{E}\widehat{r}_4(\chi,y)| = A_1 + A_2 + A_3, \tag{11.58}$$

where

$$A_1 = \frac{1}{\widehat{r}_1(\chi)} \sup_{y\in S} |\widehat{r}_4(\chi,y) - \widehat{r}_4(\chi,t_y)|,$$

$$A_2 = \frac{1}{\widehat{r}_1(\chi)} \sup_{y\in S} |\widehat{r}_4(\chi,t_y) - \mathbb{E}\widehat{r}_4(\chi,t_y)|,$$

$$A_3 = \frac{1}{\widehat{r}_1(\chi)} \sup_{y\in S} |\mathbb{E}\widehat{r}_4(\chi,t_y) - \mathbb{E}\widehat{r}_4(\chi,y)|.$$

- The treatment of A_1 is not affected by the dependence, and so (6.43) is valid.
- The treatment of A_3, can be done by using (6.44) together with (11.9) and Proposition A.5-i, and finally (6.45) is still true.
- To treat the term A_2, we use the decomposition (6.46), and because the r.r.v. U_i still satisfy the boundedness condition (6.47), we are in position to apply the Proposition A.11-ii. Putting either $b = a$ under the arithmetic condition (11.54) or $b = +\infty$ under the geometric one (11.55), and applying Proposition A.11-ii with $r = (\log n)^2$, we have:

$$P\left(\sup_{y\in S} |\widehat{r}_4(\chi,t_y) - \mathbb{E}\widehat{r}_4(\chi,t_y)| > \epsilon\sqrt{n^{-2}s_n^2 \log n}\right)$$

$$\leq Cn^{2\varsigma}\left\{\left(1+\frac{\epsilon^2}{\log n}\right)^{-\frac{(\log n)^2}{2}} + n(\log n)^{-2}\left(\frac{\sqrt{\log n}}{\epsilon s_n}\right)^{b+1}\right\}$$

$$\leq Cn^{2\varsigma}\left\{e^{-\frac{\epsilon\log n}{2}} + n(\log n)^{-2+\frac{1+b}{2}}s_n^{-b-1}\epsilon^{-b-1}\right\}$$

$$\leq C\left\{n^{2\varsigma-\frac{\epsilon^2}{2}} + n^{2\varsigma+1}s_n^{-b-1}(\log n)^{-2+\frac{1+b}{2}}\right\}.$$

The condition on s_n (see either (11.54) or (11.55)), allows us to see that for ϵ large enough there exists some $\eta > 0$ such that:

$$P\left(\sup_{y\in S}|\widehat{r}_4(\chi,t_y) - \mathbb{E}\widehat{r}_4(\chi,t_y)| > \epsilon\sqrt{n^{-2}s_n^2\,\log n}\right) = O\left(n^{-1-\eta}\right),$$

which is a stronger result than the following one:

$$\lim_{n\to\infty}\sup_{y\in S}|\widehat{r}_4(\chi,t_y) - \mathbb{E}\widehat{r}_4(\chi,t_y)| = 0,\ \ a.co.$$

By combining this result together with (11.9) and Proposition A.5-i, we get

$$\lim_{n\to\infty} A_2 = 0,\ a.co. \tag{11.59}$$

Finally, the claimed result (11.57) follows from (11.58), (6.43), (6.45) and (11.59).□

Lemma 11.17. *Under the conditions of Theorem 11.15-ii, we have for any compact $S \subset \mathbb{R}$ and for any $l = 1,\ldots,j$:*

$$sup_{y\in S}|F_Y^{\chi(l)}(y) - \widehat{F}_Y^{\chi(l)}(y)| = O\left(h^\beta + g^{\beta_0}\right) +$$

$$O_{a.co.}\left(\sqrt{\frac{s_{n,3,l}^2\,\log n}{n^2}}\right).$$

Proof. •   This proof follows the same lines as when the data are independent (see Lemma 6.17 before). Moreover, most of the steps of the proof of
Lemma 6.17 are not affected by the dependence of the data. Precisely, the
results (6.83) and (6.101) remain true. In an other hand, the denominators
appearing in (6.83) can be treated directly by mean of (11.9) and Proposition A.5-i. Finally, this proof will be complete as long as the following
result can be checked:

$$\frac{1}{\widehat{r}_1(\chi)}\sup_{y\in S}|\widehat{r}_3^{(l)}(\chi,y) - \mathbb{E}\widehat{r}_3^{(l)}(\chi,y)| = O_{a.co.}\left(\left(\frac{s_{n,3,l}\,\log n}{n^2}\right)^{\frac{1}{2}}\right). \tag{11.60}$$

•   Let us now prove that (11.60) holds. Using the same decomposition as
in the proof of Lemma 6.17 we can write that $S \subset \bigcup_{k=1}^{z_n} S_k$ where $S_k = (t_k - l_n,\ t_k + l_n)$ and where l_n and z_n can be chosen such that $l_n = Cz_n^{-1} \sim Cn^{-(l+1)\varsigma-1/2}$. Taking $t_y = \arg\min_{t\in\{t_1,\ldots,t_{z_n}\}}|y - t|$, we have:

$$\frac{1}{\widehat{r}_1(\chi)}\sup_{y\in S}|\widehat{r}_3^{(l)}(\chi,y) - \mathbb{E}\widehat{r}_3^{(l)}(\chi,y)| = B_1 + B_2 + B_3, \tag{11.61}$$

where

$$B_1 = \frac{1}{\widehat{r}_1(\chi)}\sup_{y\in S}\left|\widehat{r}_3^{(l)}(\chi,y) - \widehat{r}_3^{(l)}(\chi,t_y)\right|,$$

$$B_2 = \frac{1}{\widehat{r}_1(\chi)} \sup_{y \in S} \left| \widehat{r}_3^{(l)}(\chi, t_y) - \mathbb{E}\widehat{r}_3^{(l)}(\chi, t_y) \right|,$$

$$B_3 = \frac{1}{\widehat{r}_1(\chi)} \sup_{y \in S} \left| \mathbb{E}\widehat{r}_3^{(l)}(\chi, t_y) - \mathbb{E}\widehat{r}_3^{(l)}(\chi, y) \right|.$$

- To treat the term B_1, note that the proof of (6.105) only uses deterministic arguments, and is therefore not affected by the dependence. We get from (6.105), together with the last part of (6.31) and with the condition on $s_{n,3,l}$ (see either (11.54) or (11.55)) that:

$$B_1 = O\left(\frac{1}{\sqrt{n}}\right) = O\left(\sqrt{\frac{s_{n,3,l}^2 \log n}{n^2}}\right). \tag{11.62}$$

- Similarly, we can treat the term B_3 by combining (6.105) together with (11.9) and Proposition A.5-i. We see directly that

$$B_3 = O_{a.co.}\left(\sqrt{\frac{s_{n,3,l}^2 \log n}{n^2}}\right). \tag{11.63}$$

- It remains to treat the term B_2, that will indeed be the only one to be affected by the dependence on the variables. Note that we have for any $\epsilon > 0$:

$$P\left(\sup_{y \in S} \left| \widehat{r}_3^{(l)}(\chi, t_y) - \mathbb{E}\widehat{r}_3^{(l)}(\chi, t_y) \right| > \epsilon \right)$$

$$\leq z_n \max_{j=1...z_n} P\left(\left| \widehat{r}_3^{(l)}(\chi, t_j) - \mathbb{E}\widehat{r}_3^{(l)}(\chi, t_j) \right| > \epsilon \right).$$

Because the bounds obtained in (6.88) and (6.89) are not affected by the dependence structure, we are in position to apply the Fuk-Nagaev exponential inequality for bounded mixing variables (see Proposition A.11-ii). Applying this inequality with $r = (\log n)^2$ and either $b = a$ under the arithmetic condition (11.54) or $b = 0$ under the geometric one (11.55), we get for any t_j:

$$P\left(\left| \widehat{r}_3^{(l)}(\chi, t_j) - \mathbb{E}\widehat{r}_3^{(l)}(\chi, t_j) \right| > \epsilon \sqrt{\frac{s_{n,3,l} \log n}{n^2}} \right)$$

$$\leq C\left\{ \left(1 + \frac{\epsilon^2}{\log n}\right)^{-\frac{(\log n)^2}{2}} + n (\log n)^{-2} \left(\frac{\sqrt{\log n}}{\epsilon s_{n,3,l}}\right)^{b+1} \right\}$$

$$\leq C\left\{ e^{-\frac{\epsilon \log n}{2}} + n (\log n)^{-2+\frac{1+b}{2}} s_{n,3,l}^{-b-1} \epsilon^{-b-1} \right\}$$

$$\leq C\left\{ n^{-\frac{\epsilon^2}{2}} + n s_{n,3,l}^{-b-1} (\log n)^{-2+\frac{1+b}{2}} \right\}.$$

Both of these last results lead directly, for ϵ large enough, to:

$$P\left(\sup_{y \in S} |\widehat{r}_3^{(l)}(\chi, t_y) - \mathbb{E}\widehat{r}_3^{(l)}(\chi, t_y)| > \epsilon\sqrt{\frac{s_{n,3,l}\log n}{n^2}}\right)$$

$$\leq C\left\{n^{2\zeta - \frac{\epsilon^2}{2}} + n^{1+\zeta}s_{n,3,l}^{-b-1}(\log n)^{-2+\frac{1+b}{2}}\right\}$$

$$= O\left(n^{-1-\eta}\right),$$

where η is some real number $\eta > 0$. Note that the last inequality has been obtained from the condition on $s_{n,3,l}$ (see either (11.54) or (11.55)). We arrive finally at:

$$B_3 = O_{a.co.}\left(\sqrt{\frac{s_{n,3,l}^2\log n}{n^2}}\right). \tag{11.64}$$

This proof can be finished just by combining (11.61), (6.106), (6.107) and (11.64).

The general result given in Theorem 11.15-ii can be formulated in several different specific ways, according the knowledge we have about the covariance term s_n^2. We will only present in Corollaries 11.18 and 11.19 below two special cases for which the covariance terms have the same behavior as in i.i.d. setting. More general formulations can be found in [FLV05b]. The conditions are similar to those appearing in regression and quantile estimation (see Remark 11.2). Precisely, we will assume that

$$\exists \epsilon_1 \in (0,1), \, 0 < \psi_\chi(h) = O\left(\varphi_\chi(h)^{1+\epsilon_1}\right), \tag{11.65}$$

$$\exists \epsilon_2 \in (0,1), \, g\varphi_\chi(h) = O\left(n^{-\epsilon_2}\right), \tag{11.66}$$

and that for any $i \neq j$:

The conditional density $f_{i,j}$ of (Y_i, Y_j)

given $(\boldsymbol{\mathcal{X}}_i, \boldsymbol{\mathcal{X}}_j)$ exists and is bounded. $\tag{11.67}$

11.4.3 Application to the Geometrically Mixing Case

The next result will be a direct consequence of the properties obtained for the covariance sums in Lemma 11.11 above. This result states that, under quite unrestrictive additional assumptions, the rates of convergence for the functional kernel conditional mode estimate are the same for geometric mixing variables as for i.i.d. ones.

> **Corollary 11.18.** *Assume that the sequence $(\mathcal{X}_i, Y_i)_{i=1,\ldots,n}$ is geometrically mixing. Assume that the conditions of Theorem 6.16 hold together with (11.65), (11.66) and (11.67). Then we have:*
>
> $$\widehat{\theta}(\chi) - \theta(\chi) = O\left(\left(h^\beta + g^\beta\right)^{\frac{1}{j}}\right) + O_{a.co.}\left(\left(\frac{\log n}{ng\varphi_\chi(h)}\right)^{\frac{1}{2j}}\right). \qquad (11.68)$$

Proof. By combining the results of Lemma 11.17 and Lemma 11.11, we arrive directly at the following result:

$$sup_{y \in S}|F_Y^{\chi^{(l)}}(y) - \widehat{F}_Y^{\chi^{(l)}}(y)| = O\left(h^\beta + g^{\beta_0}\right) +$$

$$O_{a.co.}\left(\left(\frac{\log n}{ng^{2l-1}\varphi_\chi(h)}\right)^{\frac{1}{2}}\right).$$

The proof of Corollary 11.18 follows directly from this last result (applied with $l = 1$ and $S = (\theta(\chi) - \xi, \theta(\chi) + \xi)$) together with (6.98).$\square$

11.4.4 Application to the Arithmetically Mixing Case

In the following, we will show that the same rates of convergence can be achieved under an arithmetic mixing condition of sufficiently high order. Recall that ϵ_1 and ϵ_2 are defined by (11.65) and (11.66), and assume that

$$a > \frac{1 + \epsilon_1}{\epsilon_1 \epsilon_2}. \qquad (11.69)$$

> **Corollary 11.19.** *Assume that the sequence $(\mathcal{X}_i, Y_i)_{i=1,\ldots,n}$ is arithmetically mixing with order a satisfying (11.69). Assume that the conditions of Theorem 6.16 hold together with (11.65), (11.66) and (11.67). Then we have:*
>
> $$\widehat{\theta}(\chi) - \theta(\chi) = O\left(\left(h^\beta + g^\beta\right)^{\frac{1}{j}}\right) + O_{a.co.}\left(\left(\frac{\log n}{ng\varphi_\chi(h)}\right)^{\frac{1}{2j}}\right). \qquad (11.70)$$

Proof. By combining the results of Lemma 11.17 and Lemma 11.13, we arrive directly at the following result which is an extension to arithmetically mixing variables of the Lemma 6.17 stated above in the i.i.d. setting:

$$sup_{y \in S}|F_Y^{\chi^{(l)}}(y) - \widehat{F}_Y^{\chi^{(l)}}(y)| = O\left(h^\beta + g^{\beta_0}\right) +$$

$$O_{a.co.}\left(\left(\frac{\log n}{ng^{2l-1}\varphi_\chi(h)}\right)^{\frac{1}{2}}\right).$$

The proof of Corollary 11.18 follows directly from this last result (applied with $l = 1$ and $S = (\theta(\chi) - \xi, \theta(\chi) + \xi)$) together with (6.98).$\square$

11.5 Complements on Conditional Distribution Estimation

11.5.1 Convergence Results

Throughout previous sections, when dealing with conditional quantile and conditional mode, several auxilliary lemmas have been proved. These lemmas concerned asymptotic properties of the kernel conditional c.d.f. and kernel conditional density estimates. Because they could be useful by themselves and not only for quantile or mode estimation, we have decided to devote this short section to a synthetic presentation of these results. Proposition 11.20 presents almost complete convergence results of the kernel conditional cd.f. estimate (both pointwisely and uniformly over a compact set), while Proposition 11.21 will do the same for the kernel conditional density estimate. In Section 11.5.2 the rates of convergence will be specified.

Proposition 11.20. *i) Under the conditions of Theorem 11.7-i, we have for any fixed real point y:*

$$\lim_{n \to \infty} F_Y^\chi(y) = \widehat{F}_Y^\chi(y), \quad a.co.$$

ii) If in addition the bandwidth g satisfies for some $\zeta > 0$ the condition $\lim_{n \to \infty} g n^\zeta = \infty$, then for any compact subset $S \subset \mathbb{R}$ we have:

$$\lim_{n \to \infty} \sup_{y \in S} |F_Y^\chi(y) - \widehat{F}_Y^\chi(y)| = 0, \quad a.co.$$

Proof. Let us use the same notation as those used before in the proof of Proposition 6.53 or along the proof of Lemma 11.8.

i) This result has already been proved in Lemma 11.8 before.
ii) By using the same steps as to prove Proposition 6.53-ii, and by noting that the result (6.55) is still true (since it concerns deterministic terms that are not affected by the new dependence structure), it turns out that the only thing to prove is that:

$$\lim_{n \to \infty} sup_{y \in S} |\widehat{r}_3(\chi, y) - \mathbb{E}\widehat{r}_3(\chi, y)| = 0, \quad a.co.,$$

Using the compactness of S, we can write that $S \subset \bigcup_{k=1}^{z_n} S_k$ where $S_k = (t_k - l_n, \ t_k + l_n)$ and where l_n and z_n can be chosen such that:

$$l_n = C z_n^{-1} \sim C n^{-\zeta}. \tag{11.71}$$

Taking $t_y = \arg \min_{t \in \{t_1, \dots, t_{z_n}\}} |y - t|$, we have

$$\frac{1}{\widehat{r}_1(\chi)} \sup_{y \in S} |\widehat{r}_3(\chi, y) - \mathbb{E}\widehat{r}_3(\chi, y)| = D_1 + D_2 + D_3,$$

where

$$D_1 = \frac{1}{\widehat{r}_1(\chi)} \sup_{y \in S} |\widehat{r}_3(\chi, y) - \widehat{r}_3(\chi, t_y)|,$$

$$D_2 = \frac{1}{\widehat{r}_1(\chi)} \sup_{y \in S} |\widehat{r}_3(\chi, t_y) - \mathbb{E}\widehat{r}_3(\chi, t_y)|,$$

$$D_3 = \frac{1}{\widehat{r}_1(\chi)} \sup_{y \in S} |\mathbb{E}\widehat{r}_3(\chi, t_y) - \mathbb{E}\widehat{r}_3(\chi, y)|.$$

- The treatment of D_1 is not affected by the dependence structure. So, the result (6.60) remains true.
- The treatment of D_3 can be done by using (6.59) together with (11.9) and Proposition A.5-i, and finally (6.61) is still true.
- To treat the term D_2, we use the decomposition (6.46), and because the r.r.v. $\Gamma_i(y)$ and Δ_i are bounded (since the kernels H and K are bounded), we are in position to apply Proposition A.11-ii. Putting either $b = a$ under the arithmetic condition (11.54) or $b = +\infty$ under the geometric one (11.55), and applying Proposition A.11-ii with $r = (\log n)^2$, we have:

$$P(\sup_{y \in S} |\widehat{r}_3(\chi, t_y) - \mathbb{E}\widehat{r}_3(\chi, t_y)| > \epsilon \sqrt{n^{-2} s_{n,3,0}^2 \log n})$$

$$\leq C n^{\varsigma} \left\{ \left(1 + \frac{\epsilon^2}{\log n}\right)^{-\frac{(\log n)^2}{2}} + n (\log n)^{-2} \left(\frac{\sqrt{\log n}}{\epsilon s_{n,3,0}}\right)^{b+1} \right\}$$

$$\leq C n^{\varsigma} \left\{ e^{-\frac{\epsilon \log n}{2}} + n (\log n)^{-2 + \frac{1+b}{2}} s_{n,3,0}^{-b-1} \epsilon^{-b-1} \right\}$$

$$\leq C \left\{ n^{\varsigma - \frac{\epsilon^2}{2}} + n^{\varsigma+1} s_{n,3,0}^{-b-1} (\log n)^{-2 + \frac{1+b}{2}} \right\}.$$

The condition on $s_{n,3,0}$ (see either (11.31) or (11.32)), allows us to see that for ϵ large enough there exists some $\eta > 0$ such that:

$$P\left(\sup_{y \in S} |\widehat{r}_3(\chi, t_y) - \mathbb{E}\widehat{r}_3(\chi, t_y)| > \epsilon \sqrt{n^{-2} s_{n,3,0}^2 \log n}\right)$$

$$= O\left(n^{-1-\eta}\right), \qquad (11.72)$$

which is a stronger result than the following one:

$$\lim_{n \to \infty} \sup_{y \in S} |\widehat{r}_3(\chi, t_y) - \mathbb{E}\widehat{r}_3(\chi, t_y)| = 0, \ a.co.$$

By combining this result, together with (11.9) and Proposition A.5-i, we get

$$\lim_{n \to \infty} A_2 = 0, \ a.co.$$

This is enough to complete our proof.□

Proposition 11.21. *i) Under the conditions of Theorem 11.15-i we have for any fixed real number y:*

$$\lim_{n\to\infty} f^X_Y(y) = \widehat{f}^X_Y(y), \quad a.co.$$

ii) If in addition (6.31) holds, then we have for any compact $S \subset \mathbb{R}$:

$$\lim_{n\to\infty} \sup_{y\in S} |f^X_Y(y) - \widehat{f}^X_Y(y)| = 0, \quad a.co.$$

Proof. i) This result is a special case, with $l = 1$, of Lemma 11.10 above.
ii) This result was already stated before in Lemma 11.16.

11.5.2 Rates of Convergence

The rates of convergence are stated precisely in both of the next propositions. We start with rates of almost complete convergence for the functional kernel c.d.f. estimate. In Proposition 11.22 below we just present a general result in which the rate of convergence is expressed as a function of the covariance terms of the estimate. By combining Proposition 11.22 together with the asymptotic bounds given for these covariance terms in Section 11.3.3 (resp. in Section 11.3.4) we will directly provide much more explicit rates of convergence for geometrically (resp. for arithmetically) mixing processes.

Proposition 11.22. *i) Under the conditions of Theorem 11.7-ii, we have for any fixed real number y:*

$$F^X_Y(y) - \widehat{F}^X_Y(y) = O\left(h^\beta + g^\beta\right)$$
$$+ O_{a.co.}\left(\frac{\sqrt{(\max\{s_{n,3,0}, s_{n,1}\})^2 \log n}}{n}\right).$$

ii) If in any adddition the bandwidth g satisfies for some $a > 0$ the condition $\lim_{n\to\infty} gn^a = \infty$, then for any compact subset $S \subset \mathbb{R}$ we have:

$$\sup_{y\in S} |F^X_Y(y) - \widehat{F}^X_Y(y)| = O\left(h^\beta + g^\beta\right)$$
$$+ O_{a.co.}\left(\frac{\sqrt{(\max\{s_{n,3,0}, s_{n,1}\})^2 \log n}}{n}\right).$$

Proof. The result i) was already stated by Lemma 11.9. It remains just to prove ii). The proof is performed over the same step as for Proposition 6.19-ii. Note that the result (6.113) is not stochastic, and so it remains valid under the new dependent setting. Moreover, it follows from (11.9), Proposition A.5-i and (11.72) that:

$$\frac{1}{\widehat{r}_1(\chi)} \sup_{y \in S} |\widehat{r}_3(\chi, y) - \mathbb{E}\widehat{r}_3(\chi, y)| = O_{a.co.} \left(\frac{\sqrt{s_{n,3,0}^2 \log n}}{n} \right).$$

This last result together with (6.113), (6.18), (11.9) and Proposition A.5-i is enough to prove our result.□

Now we present the rate of convergence of the functional kernel density estimate. The rate is expressed in a general way. Note that the results of Proposition 11.23 below can be combined with the asymptotic bounds given in Section 11.4.3 (resp. in Section 11.4.4) to get directly much more explicit rates of convergence for geometrically (resp. for arithmetically) mixing processes.

Proposition 11.23. *i) Under the conditions of Theorem 11.15-ii we have for any fixed real number y:*

$$f_Y^X(y) - \widehat{f}_Y^X(y) = O\left(h^\beta + g^\beta \right)$$

$$+ O_{a.co.} \left(\frac{\sqrt{(\max\{s_{n,4}, s_{n,1}\})^2 \log n}}{n} \right).$$

ii) If in addition (6.31) holds, then we have for any compact subset $S \subset \mathbb{R}$:

$$\sup_{y \in S} |f_Y^X(y) - \widehat{f}_Y^X(y)| = O\left(h^\beta + g^\beta \right)$$

$$+ O_{a.co.} \left(\frac{\sqrt{(\max\{s_{n,4}, s_{n,1}\})^2 \log n}}{n} \right).$$

Proof. This proposition is just a special case (with $l = 1$ and $\beta = \beta_0$) of results that have already been proved along the previous calculations (see the second part of Lemma 11.10 and Lemma 11.17). □

11.6 Nonparametric Discrimination of Dependent Curves

11.6.1 Introduction and Notation

In this section we will show how the kernel nonparametric methodology developed in Chapter 8 to discriminate a set of independant functional data behaves for dependent ones. We will concentrate on theoretical supports for this functional curves discrimination problem, equivalently known as a supervised curves classification problem. More precisely, we will state extensions to mixing functional variables of the theorems stated in Section 8.5. As described before in Chapter 8 for i.i.d. variables, functional kernel discrimination can be seen as a rather direct application of the functional kernel regression methodology. This will also be the case here, and we will see that our proofs will be obtained by quite direct applications of the results obtained in Section 11.2 for kernel regression with mixing functional variables.

As presented before (see Section 8.2 for more details), the discrimination (or supervised classification) statistical problem involves a sample $(\boldsymbol{X}_i, Y_i)_{i=1,\dots,n}$ of n pairs, each having the same distribution as a pair $(\boldsymbol{X}, Y)$. In the functional setting $\boldsymbol{X}$ is a f.r.v. (more precisely, it takes values into some semi-metric vector space (E, d)), and Y is a categorical response taking values into some finite set $\overline{G} = \{1, \dots, G\}$. Let χ denotes a fixed element in E. The discimination problem consists in predicting in which class, among the G ones, belongs this new functional element χ. The way to do that consists in estimating all the G posterior probabilities:

$$p_g(\chi) \;=\; P\left(Y = g | \boldsymbol{X} = \chi\right), \; \forall g \in \overline{G},$$

and to assign χ to the class $\hat{y}(\chi)$ having the highest estimated posterior probability:

$$\hat{y}(\chi) \;=\; \arg\max_{g \in \overline{G}} \widehat{p}_g(\chi).$$

As motivated in Section 8.2, the estimation of p_g can be carried out by means of functional kernel ideas. Precisely, we define

$$\widehat{p}_g(\chi) \;=\; \widehat{p}_{g,h}(\chi) \;=\; \frac{\sum_{i=1}^{n} 1_{[Y_i=g]} \, K\left(h^{-1} d(\chi, \boldsymbol{X}_i)\right)}{\sum_{i=1}^{n} K\left(h^{-1} d(\chi, \boldsymbol{X}_i)\right)}, \tag{11.73}$$

where K is an asymmetrical kernel (see Section 4.1.2 and Definition 4.1) and h is the bandwidth (a strictly positive smoothing parameter). The aim is to state theoretical properties for this kernel functional posterior probability estimate in the situation where the data $(\boldsymbol{X}_i, Y_i)_{i=1,\dots,n}$ are not independent but assumed to satisfy some strong mixing condition.

11.6.2 Complete Convergence Properties

In the next theorem we will state the almost complete convergence properties of the estimate $\widehat{p}_g$ without (resp. with) rate under a continuity-type model (resp. under a Lipschitz-type model) for the posterior probability nonlinear operator $p_g(.)$. This theorem is an extension of the results presented in Section 8.5, in the sense that we do not assume now that the data $(\boldsymbol{X}_i, Y_i)_{i=1,...,n}$ are independent. One can expect the behaviour of the estimate to be linked with the dependence between the sample pairs. This will be seen clearly in the next results through the role played by the following covariance parameters on the rate of convergence. These covariance parameters are defined by:

$$s_{n,5,g} = \sum_{i=1}^{n} \sum_{j=1}^{n} cov(1_{[Y_i=g]}\Delta_i, 1_{[Y_j=g]}\Delta_j), \qquad (11.74)$$

$$s_{n,1} = \sum_{i=1}^{n} \sum_{j=1}^{n} cov(\Delta_i, \Delta_j), \qquad (11.75)$$

$$s_{n,g} = max\{s_{n,1}, s_{n,5,g}\}, \qquad (11.76)$$

where

$$\Delta_i = \frac{K\left(h^{-1}d(\chi, \boldsymbol{X}_i)\right)}{\mathbb{E}\, K\left(h^{-1}d(\chi, \boldsymbol{X}_1)\right)}.$$

In comparison with the i.i.d. results given in Theorems 8.1 and 8.2 some assumptions are needed to control the covariance effects. Precisely, we will assume either that:

$$\left\{ \begin{array}{c} (\boldsymbol{X}_i, Y_i)_{i=1,...,n} \text{ are strongly mixing} \\ \text{with arithmetic coefficients of order } a > 1, \text{ and} \\ \exists \theta > 2, s_{n,g}^{-(a+1)} = o(n^{-\theta}), \end{array} \right. \qquad (11.77)$$

or that

$$\left\{ \begin{array}{c} (\boldsymbol{X}_i, Y_i)_{i=1,...,n} \text{ are strongly mixing} \\ \text{with geometric coefficients, and} \\ \exists \theta > 1, s_{n,g}^{-1} = o(n^{-\theta}). \end{array} \right. \qquad (11.78)$$

Theorem 11.24. *Put* $s_{n,g} = max\{s_{n,1}, s_{n,5,g}\}$, *and assume that either (11.77) or (11.78) is satisfied. Let g be fixed in $\{1, \ldots, G\}$. Then :*
i) Under the conditions of Theorem 8.1 we have:

$$\lim_{n\to\infty} \widehat{p}_{g,h}(\chi) = p_g(\chi), \quad a.co.,$$

ii) Under the conditions of Theorem 8.2 we have:

$$\widehat{p}_{g,h}(\chi) - p_g(\chi) = O\left(h^\beta\right) + O_{a.co.}\left(\frac{\sqrt{s_n^2 \log n}}{n}\right).$$

Proof. Both results i) and ii) can be obtained by noting that we have:

$$E\left(1^{m}_{[Y=g]}|\boldsymbol{X}=\chi\right) \;=\; P(Y=g|\boldsymbol{X}=\chi) \;\overset{def}{=}\; p_g(\chi), \qquad (11.79)$$

in such a way that, for each g, the estimation of the operator p_g can be seen as a special case of the estimation of the regression operator of the variable $1_{[Y=g]}$ given $\boldsymbol{X}$. The boundedness of the variable $1_{[Y=g]}$ insures that the condition (6.4) is satisfied, and so we are in position to apply Theorem 11.1. This is enough to complete this proof.□

The rate of convergence in Theorem 11.24-ii is stated in a general way as a function of the covariance term s_n^2. Under some specific assumptions on the mixing coefficients we can get much more explicit results. To fix the ideas, we present below two corollaries for which this rate of convergence can be shown to be the same as in Theorem 8.2 for i.i.d. variables.

Corollary 11.25. *Assume that the sequence $(\boldsymbol{X}_i, Y_i)_{i=1,\dots,n}$ is geometrically mixing. Assume that the conditions of Theorem 8.2 hold together with (11.11) and (11.12). Then we have:*

$$\widehat{p}_{g,h}(\chi) - p_g(\chi) \;=\; O\left(h^\beta\right) + O_{a.co.}\left(\sqrt{\frac{\log n}{n\varphi_\chi(h)}}\right).$$

Proof. Once again, because of (11.79) the estimation of p_g is a regression estimation problem. Note that the condition (11.10) is obviously satisfied by the response variable $1_{[Y=g]}$, in such a way that Lemma 11.3 can be applied to get:

$$s_{n,g}^2 \;=\; O\left(\frac{n}{\varphi_\chi(h)}\right). \qquad (11.80)$$

This last result, combined with the result of Theorem 11.24-ii is enough to get the claimed result. □

Corollary 11.26. *Assume that the sequence $(\boldsymbol{X}_i, Y_i)_{i=1,\dots,n}$ is arithmetically mixing with order a satisfying (11.20). Assume that the conditions of Theorem 8.2 hold together with (11.11) and (11.12). Then we have:*

$$\widehat{p}_{g,h}(\chi) - p_g(\chi) \;=\; O\left(h^\beta\right) + O_{a.co.}\left(\sqrt{\frac{\log n}{n\varphi_\chi(h)}}\right).$$

Proof. The proof follows exactly the same steps as for Corollary 11.25 before, except that the result (11.80) is obtained by using Lemma 11.5 rather than Lemma 11.3. □

11.7 Discussion

11.7.1 Bibliography

The nonparametric modelling of functional dependent data is a very recent field of research, and it turns out that there is only a little literature on the topic. As far as we know, the first paper in this direction was provided by [FGV02] in regression setting. The literature in nonparametric prediction from functional dependent variables involves only [FV04] and [M05] for regression methods, [FRV05] for conditional quantile methods and [FLV05b] for conditional mode methods. Concerning nonparametric classification for dependent functional variables, the only paper attacking the supervised classification problem can be found in [FV04]. Hovewer, in spite of this very small existing bibliography, we support the idea that most of the nonparametric methodologies developed for i.i.d. functional variables can be extended to dependent functional samples, pending of course some suitable adaptations. We hope that the contents of this book will help in convincing people to share with us this point of view, since (except for the unsupervised classification problem) all the results presented in Parts II and III of the book have been extended to dependent situations.

11.7.2 Back to Finite Dimensional Setting

In finite dimensional setting, nonparametric statistics for dependent variables have been extensively studied. Even restricting the purpose to kernel estimation and to mixing dependent processes, the list is too wide to make an exhaustive bibliographical survey here (and this is particularly true concerning regression problems). In order to have a chronological view of how the knowledge on kernel regression estimation for mixing processes grew up, some arbitrarily selected set of references would be [R83], [C84], [CH86], [GHSV89], [Ro90], [V91], [Y94b], [BF95b], [L96], [B98], [L99], [KD02], [KSK04]. Of course, most of the results presented in these papers can be directly aplied to derive analogous results in nonparametric kernel discrimination of dependent variables. In conditional kernel c.d.f. and quantiles estimation a sample of references would be [ZL85], [G90], [V91], [Y93c], [WZ99], [C02] and [GSY03], while in kernel conditional density and mode it would be [CHH87], [M89], [V91], [C91], [X94], [O97], [QV97], [LO99], [IM02], [GSY03] and [deGZ03].

Exactly as was the case for i.i.d. problems (see the discussion in Section 6.4.2), even if the main goal of this book was to discuss recent advances in functional settings it is worth noting that all the results presented in this chapter are written in the general setting when the explanatory variable takes value in some abstract semi-metric space (E, d). That means that all the results presented before can be directly applied to the classical un-functional setting by taking $E = \mathbb{R}^p$ and by chosing d to be the usual euclidian metric. Therefore, as by-products of the infinite dimensional methodologies, we also get interesting new contributions in the classical un-functional setting. This is particularly true for the nonparametric discrimination which did not receive much theoretical attention in the past for dependent data, even in the finite dimensional setting. As discussed in Section 6.4.2 for i.i.d. variables, the most important point here is to note that the direction followed in this book is based on considerations on small ball probabilities of the explanatory variables, without any need for introducing the density function of this variable. On the other hand, all the usual literature presented above assumes the existence (and smoothness) of such a density function. In other words, as direct consequences of the infinite dimensional results we get new contributions in classical un-functional nonparametric kernel regression, conditional quantiles and conditional mode estimation from mixing samples. The reader will find in Section 13.4 (respectively in Section 13.5) general considerations presenting how such a one-dimensional (respectively a p-dimensional) application of our infinite dimensional results can be easily carried out with interesting effects.

11.7.3 Some Open Problems

Of course, the same open questions as those discussed in Section 6.4.3 remain open in the dependent situation. Among all the open questions proposed in Section 6.4.3, we wish however to emphasize both of the following which are evidently the most determinant:

Open question 10 : Bandwidth choice. *How can automatical bandwidth selection procedures for dependent data be developed?*

and

Open question 11 : Semi-metric choice. *How can we choose the semi-metric in practice?*

Until now, none of them have been theoretically attacked in the literature? Concerning bandwidth selection, the dependence among the data will play an important role, and possible answers should rest on suitable adapation of what exists in the finite dimensional nonparametric literature on this point (see [HV90], [HV92], [V93], [C94], [BF95], [HLP95], [C95], [H96], [RT97] and [S01] for a non-exhaustive list of references). Our guess is that the cross-validation procedure (see [HV92]) could be the easiest one to be adapted to the functional setting. Concerning the semi-metric selection problem, this turns to be not specific to dependent data but it is rather a common feature of all the functional methodologies developped all along the book. This question is investigated specifically in Chapter 13.

We will not take space here again for the other open questions discussed in Section 6.4.3, because they can be attacked by following the same general ideas as proposed in Section 6.4.3. Let us just mention two recent works, both on dependent functional regression, which state some first advances on some of these problems. The first one is the contribution by [M05] which, by following the same kind of proofs as those of this book, completes the asymptotic study by stating a limit gaussian distribution. The second one is by [AFK05] and studies a single index model for functional dependent regression problems.

Finally, let us emphasize a new open problem which is peculiar to the dependent setting, and which is linked with the dependence sructure itself. The question would be to introduce other kinds of dependence structures in order to cover wider situations than those that are allowed with mixing processes. In particular, there are some advances in finite dimensional settings (see for instance [HH090] or [EV03]) in which long memory dependence structures are introduced, still keeping a completely nonparametric framework (that is, without any distribution assumption such as gaussianity for instance). Clearly, such an approach in our functional context would be greatly interesting, and we hope that the ideas developed in these two papers could be adapted to the infinite dimension in order to bring some contribution to the following question:

Open question 12 : On the long memory assumption. *Is it possible to include long dependence structures in the nonparametric functional context?*

12

Application to Continuous Time Processes Prediction

In Chapter 11, the functional nonparametric methodology was shown to have appealing theoretical supports for dependent statistical samples. The aim of this chapter is to show how this methodology can be used in practical situations for analysing time series. After a short discussion in Section 12.1 on how nonparametric finite dimensional statistics are used in the standard literature to treat time series, Section 12.2 explains how time series analysis can be viewed as specific functional nonparametric problems for dependent data for which all the methodology described in Chapter 11 will apply directly. Then we will see in Section 12.3 that, despite their rather technical look, these nonparametric functional methods are easy to implement. To emphasize this point, a real dataset will be quickly treated in Section 12.4. Finally, the source codes, functional datasets, descriptions of the $R/S+$ routines, guidelines and examples of use are detailed in the companion website *http://www.lsp.ups-tlse.fr/staph/npfda*.

12.1 Time Series and Nonparametric Statistics

The statistical analysis of some time series $\{Z_t, t \in \mathbb{R}\}$ is always linked with models and methods involving dependent data. Let us look for instance at the standard case when the process has been observed until time T and when the problem is to predict some future value Z_{T+s} of the process. Usually, the process is observed at a grid of N discretized times, and the observations are denoted by $\{Z_1, \ldots Z_N\}$. The first step for predicting future values is to decide how much information has to be taken into account from the past?

The simpler situation consists in predicting the future just by taking into account one single past value. This is usually done by constructing some two-dimensional statistical sample of size $n = N - s$:

$$X_i = Z_i \text{ and } Y_i = Z_{i+s}, \ i = 1, \ldots, N - s, \tag{12.1}$$

in such a way that the problem turns to be a standard prediction problem of a real valued response Y given some real explanatory variable X. The only additional difficulty comes from the obvious necessity for allowing dependence structure in the statistical sample (X_i, Y_i). This approach can be used for many different statistical purposes and with various nonparametric estimates. It can lead to appealing results in practical situations as shown in several different real data studies that have been performed in the statistical literature (see for instance [HV92], [CD93], [H96], [R97], [GSY03], or [Co04] for the treatment of several time series coming from various fields of applied statistics).

Of course, this univariate modelling of the explanatory variable can be too restrictive to take into account sufficient information in the past of the series. To bypass this problem one could think in terms of constructing some $(p+1)$-dimensional statistical sample (of size $n = N - s - p + 1$) in the following manner:

$$\mathbf{X_i} = (Z_{i-p+1}, \ldots, Z_i) \text{ and } Y_i = Z_{i+s}, \ i = p, \ldots, N - s, \qquad (12.2)$$

in such a way that the problem turns to be a standard prediction problem of a real valued response Y given some p-dimensional explanatory variable $\mathbf{X}$. Once again, as before when $p = 1$, it is necessary to attack this regression problem by allowing some dependence into the statistical sample $(\mathbf{X_i}, Y_i)$. Indeed, nonparametric approach to such a multidimensional prediction problem suffers from the curse of dimensionality (see discussions and references in Sections 3 and 13.5). Because of this curse of dimensionality the question of the choice of the order p turns to be a crucial one (see for instance [V95], [V02], [GQV802], [TA94] and [AT90] for an unexhaustive list of recent approaches to this question and for more references). From a practical point of view, most people prefer the use of semi-parametric modelling in order to reduce the effects of the dimension. It is out of the scope of this book to discuss in detail these semi-parametric and/or reduction dimension modelling approaches. For that, and to stay inside within the most recent references, we could encourage the reader to look at the advances provided by [GT04], [G99], [AR99], [G98], as well as at the general discussions presented by [HMSW04], [FY03], [HLG00] and [Gh99].

Finally, if one wishes to minimize the modelling errors by staying in pure nonparametric framework, it seems that there is a trade-off to balance between taking too many explanatory past values of the series (but with bad influence on the statistical performance of the estimates) and insuring good behaviour of the estimates (but by reducing the information from the past). We will see in the next section that the functional methodology is one way to answer this question.

12.2 Functional Approach to Time Series Prediction

The functional approach to time series forecasting consists in taking as past explanatory values a whole continuous path of the process. To simplify the notation assume that $N = n\tau$ for some $n \in \mathbb{N}^*$ and some $\tau > 0$. We can build a new statistical sample of size $n - 1$ in the following way:

$$\mathcal{X}_i = \{Z(t), (i-1)\tau < t \leq i\tau)\} \text{ and } Y_i = Z(i\tau + s), \ i = 1, \ldots, n-1, \quad (12.3)$$

in such a way that the forecasting question turns to be a prediction problem of the scalar response Y given a functional variable $\mathcal{X}$. Such a problem has to be attacked without necessarily assuming independence between the statistical pairs $(\mathcal{X}_i, Y_i)$. Starting with [B91] such a functional approach of time series forecasting has been widely attacked when a parametric (linear) shape is assumed for the link between past and future values of the process (see [B00] for an extensive discussion, [G02], [B03], and [DG05] for the most recent advances, [DG02] for an environmental application, and [B05] for a basic course in this field).

Until the last few years, except for an early paper by [BD85], this problem was never investigated in a nonparametric way. The recent kernel functional dependent methodology studied in Chapter 11 allows for doing that. Indeed, the kernel methods based on the functional modelling (12.3) are really taking into account a wide part of the past of the process, and avoid by this way the drawbacks of the standard univariate modelling (12.1). On the other hand, the asymptotic results stated in Chapter 11 are showing their relative unsensitivity to the dimensionality of the problem, and this avoids the drawbacks of the standard multivariate modelling (12.2). This last point is obtained depending on a suitable semi-metric choice as discussed in detail in Chapter 13.

Note that all these considerations are independent of the statistical method that will be used. The aim of this chapter is to complete the theoretical advances provided before by some computational issues, concerning regression, as well as conditional quantile and mode approaches to the prediction problem. The ease of implementation of all the methods will be seen in Section 12.3 through the presentation of some $R/S+$ procedures. Then, a short case study based on the economic time series presented in Section 2.3 will show the good behaviour of these nonparameric functional approaches of forecasting for finite real statistical samples.

The last point to be noted is that this functional approach to time series could be easily extended by allowing for more general response values of the form:

$$\mathcal{X}_i = \{Z(t), (i-1)\tau < t \leq i\tau)\} \text{ and } Y_i = g(\mathcal{X}_{i+1}), \ i = 1, \ldots n-1, \quad (12.4)$$

where g is a real valued known function corresponding to the specific statistical problem one wishes to address. A typical example of function g is the following one:

$$g(\mathbfcal{X}_{i+1}) = \max_{i\tau < t \le (i+1)\tau} Z(t),$$

which is particularly interesting in environmetrics (see for instance [AV04] for an application to ozone peak forecasting). Of course many other choices of g are possible. When g takes continuous real values the user will have to develop standard prediction tools like regression (see Section 11.2), conditional quantile (see Section 11.3) or conditional mode (see Section 11.4). In counterpart, when g is only taking a finite number of values the curves discrimination technique (see Section 11.6) will be more accurate.

12.3 Computational Issues

The statistical forecasting techniques presented before in Chapter 11 (as well as regression, conditional quantile or conditional mode) are defined exactly by the same expressions as for independent statistical samples (see Chapter 6). The same thing applies for the functional curves discrimination methodology (compare Chapter 8 and Section 11.6). The asymptotic studies performed in Chapter 11 show that the behaviour of such functional methods remains good in time series context, with (sometimes) some changes in the rates of convergence and depending (sometimes) on additional assumptions on the smoothing parameters. However, this could be completely transparent for the user. Indeed all the routines presented before for independent samples can be directly applied in time series analysis, as soon as the data have been reorganized as indicated in (12.3) or (12.4). This will be precisely explained in the forthcoming section through some real time series data applications. The reader will find more details on all the methods available for analyzing time series by going back to previous parts of this book (see Chapter 7 and Section 8.3.3).

12.4 Forecasting Electricity Consumption

12.4.1 Presentation of the Study

This section focuses on an application to data coming from econometrics. These discretized data are the economic time series described in Section 2.3. Not to mask the main of purpose of this book, we will directly work with the differenciated log data. The data are recorded as a sequence of real numbers. Here, the dataset is composed of $N = 336$ real values $\{z_i, i = 1, \ldots, 336\}$ as displayed in Figure 2.7. First, one has to decide which past values have to be taken into account for prediction. In order to apply the functional methodology, one has to cut the original time series in a set of functional data. Here we have decided to predict future electrical consumption by using the consumption data for the whole last year. That means that, with the notation of Section 12.2, we have choosen $\tau = 12$. This way, we have constructed the functional data presented in Figure 2.8. Precisely, to apply our $R/S+$ routines, the data have to be put into a new matrix file of size 28×12, which is organized as follows:

	Col 1	$\cdots$	Col j	$\cdots$	Col 12
Row 1	z_1	$\cdots$	z_j	$\cdots$	z_{12}
$\vdots$	$\vdots$	$\vdots$	$\vdots$	$\vdots$	$\vdots$
Row i	$z_{1+12(i-1)}$	$\cdots$	$z_{j+12(i-1)}$	$\cdots$	z_{12i}
$\vdots$	$\vdots$	$\vdots$	$\vdots$	$\vdots$	$\vdots$
Row 28	z_{325}	$\cdots$	z_{324+j}	$\cdots$	z_{336}

Here, in order to illustrate our purpose, we will not use the 28th year and we will predict it by means of the data corresponding to the 27 previous ones. To use the nonparametric functional methods, one has first to decide what is the horizon of prediction that is desired (that is, with the notation introduced before, what is s). Then, for fixed s, the data will be reorganized into a functional explanatory sample $\{\chi_i, i = 1, \ldots, 26\}$ which will be loaded in the following 26×12 matrix:

z_1	$\cdots$	z_j	$\cdots$	z_{12}
$\vdots$	$\vdots$	$\vdots$	$\vdots$	$\vdots$
$z_{1+12(i-1)}$	$\cdots$	$z_{j+12(i-1)}$	$\cdots$	z_{12i}
$\vdots$	$\vdots$	$\vdots$	$\vdots$	$\vdots$
z_{301}	$\cdots$	z_{300+j}	$\cdots$	z_{312}

and a response real sample $\{y_i, i = 1, \ldots, 26\}$, which will be loaded in the following 26-dimensional vector:

z_{12+s}	$\cdots$	z_{12i+s}	$\cdots$	z_{312+s}

For fixed horizon s, we can predict the value $\widehat{z}_{324+s}$ by using any technique among the three ones which are described in Chapter 11. Our goal is not to make a full analysis of this economic dataset, and to make things clearer we will just present the results obtained with $R/S+$ routines involving automatic bandwidth choices. More precisely, each among the three $R/S+$ routines `funopare.knn.lcv`, `funopare.mode.lcv` and `funopare.quantile.lcv` have been used to compute the predicted value $\widehat{z}_{324+s}$ obtained respectively by the kernel functional estimate of the regression operator (see Section 11.2.1), by the kernel functional conditional mode estimate (see Section 11.4.1) and by the kernel functional conditional median estimation technique (see Section 11.3.1). These routines were already introduced in Chapter 7. Concerning the semi-metric chosen for the nonparametric forecasting procedures, the small number of discretization points for each curve (exactly 12) suggested the use of a semi-metric based on functional principal components ideas. Precisely, we used the PCA semi-metric d_q^{PCA} defined in Section 3.4.1, and we took the parameter q which allows us to get the best empirical mean square errors as defined in Section 7.2.1 ($q = 5$ for funopare.knn.lcv, $q = 2$ for funopare.mode.lcv and funopare.quantile.lcv).

12.4.2 The Forecasted Electrical Consumption

We recall that the $R/S+$ commandlines for obtaining the presented predictions and their corresponding explanations allowing us to load the dataset and to run the subroutine are available in the companion website[1] of this book. The predictions have been achieved for any value of $s \in \{1, \ldots 12\}$. The results of the three forecasting procedures are presented in Figure 12.1.

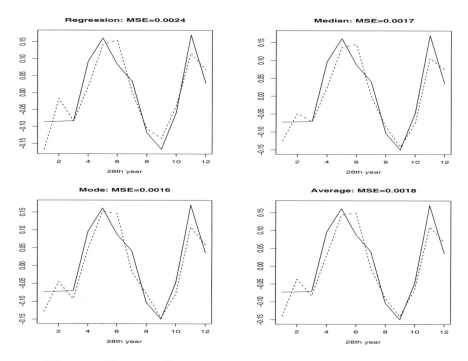

Fig. 12.1. Electricity Consumption: the Forecasting Methods in Action

---

[1] http//www.lsp.ups-tlse.fr/staph/npfda

In Figure 12.1 each of the three previous plots is concerning one among the different forecasting methods (regression, median or mode), and the dotted line (respectively the solid line) corresponds to the true observed electrical consumption (respectively to the forecasted ones) for the 28th year. The fourth plot corresponds to the forecasting that one would do by averaging the three previous ones. What could be said from these results is that each among the three functional approaches for time series forecasting gives appealing results on this dataset, with some slight advantage for the conditional median and mode forecasting methods which seem less sensitive to the high variability appearing around February (see Figure 2.8). This example was selected to show what happens when the number of discretizations for each curve is small (here it is 12) and when the statistical sample is also small (here it is $n = 26$). Of course one can expect much more precise results for larger sample sizes (see for instance [FRV05] for an application to the larger climatic El Niño time series including the construction of confidence bands). Anyway, it is worth noting that neither the dependence between the curves nor the small numbers of data obstructs the nice behaviour of the nonparametric functional methods.

12.4.3 Conclusions

Of course, this is just an example of what can be done. Readers interested in familiarizing themselves with the functional technology for time series analysis could look at how the methods behave when some parameters are changed (for instance the smoothing parameters, the order of the pca semi-metric, the kind of semi-metric, . . .), or when some other problems are attacked (such as discrimination or unsupervised classification of the yearly curves, . . .). Indeed, any of the routines previously run for independendent samples in Chapters 7 and 8 of this book remain valid in the same way for dependent samples.

On the other hand, any user could use these procedures to analyze his or her own time series. Of course, to insure the good behaviour of the procedures one will have to choose carefully the semi-metric (see discussions in Chapters 3 and 13). Several different kinds of semi-metric have already been programmed (see again Chapter 3). Moreover, the user who would need to program a new one for its own time series could incorporate it easily in the functional nonparametric procedures.

Part V

Conclusions

13

Small Ball Probabilities and Semi-metrics

13.1 Introduction

All the theoretical advances in nonparametric statistics for functional variables presented above show powerfully the key role played by the small ball probability function, both on the several different hypothesess made throughout this book and on the rates of convergence. Clearly, this function depends on the topological structure existing on the functional semi-metric space and which is induced by the semi-metric itself.

The main aim of this chapter is to describe precisely all the theoretical links existing between the small ball probability functions and the semi-metrics. In particular, we will present some examples of usual processes for which the small ball probability function can be evaluated explicitely. Purely functional examples are presented in Section 13.3, while Sections 13.4 and 13.5 will go back to standard finite dimensional ones. All these theoretical considerations will complete the empirical ideas discussed in Chapter 3 as well as the different case studies presented throughout the book which indicated that functional nonparametric methods could have quite interesting effect on real data situations if (and only if) one has selected a suitable semi-metric. Finally, we will see that in functional nonparametric statistics the semi-metric modelling turns to be the key point both for practical and theoretical issues.

As a by-product, we will see that even if infinite dimensional setting is the main purpose of this book, the general approach that we have followed here can be of interest in finite dimensional nonparametric problems. We will see in Section 13.4 how this approach allows us to extend several results existing in usual one-dimensional nonparametric statistical problems. We will also see in Section 13.5 how the approach can provide a new way to attack the curse of dimensionality in multivariate nonparametric problems.

13.2 The Role of Small Ball Probabilities

Recall that $\mathcal{X}$ is a random variable taking values into some metric-space (E, d), and that χ is a fixed (deterministic) element of E. For any of the various nonparametric problems treated earlier, each asymptotic result is directly linked with the measure (with respect to the probability distribution of $\mathcal{X}$) of a ball of center χ. It turns that the following function:

$$\varphi_\chi(.) = P(\mathcal{X} \in B(\chi, .)),$$

plays a crucial role. More precisely, the key point is the behaviour of the function $\varphi(.)$ when the radius of the ball tends to zero, and this is the reason why it is called *small ball probability function* or equivalently *concentration function*.

To fix the ideas, look for instance at the result provided in Theorem 6.11 for kernel functional regression estimation (but keep in mind that everything said here will concern equivalently all other rates of convergence given earlier in this book). Theorem 6.11 stated that, under suitable conditions, the kernel nonparametric estimate $\hat{r}$ constructed with a bandwidth h was converging to the true nonlinear regression operator r with a rate of convergence of the form:

$$O\left(h^\beta\right) + O\left(\sqrt{\frac{\log n}{n\varphi_\chi(h)}}\right).$$

While the first component comes from the bias of the estimate and depends only on the smoothness of the operator r, the second one comes from the variablity of the estimate and is therefore highly linked with the concentration of the data. The Lipschitz parameter β as defined in condition (5.12) is linked with the smoothness of r, while the small ball probability function $\varphi_\chi(h)$ directly measures the concentration of the functional variable $\mathcal{X}$. The less dispersed are the functional data $\mathcal{X}_i, \ldots \mathcal{X}_n$, the more efficient will be the estimator. With other words, the more concentrated the random variable $\mathcal{X}$, the higher will be the small ball probability function φ_χ and the faster will be the rate of convergence of the functional nonparametric estimate to the true target operator.

At this stage, it is natural to wish the small ball probability function to be as high as possible to avoid possible overdispersion effects. This probabilistic point of view can be however balanced by some considerations of the topological structure of the functional space. Indeed, the notion of concentration (and therefore the function φ_χ itself) is directly linked with the structure of the space E, in such a way that what could appear to be a purely probabilistic question turns to be primarily a topological one. Therefore, one could expect to be able to reduce the possible overdispersion effects just by changing the topological structure on the space E, that is by changing the semi-metric d.

Finally, it should be emphasized that the roles of the probability distribution of the functional variable $\mathcal{X}$ and of the semi-metric d are completely indissociable in the nonparametric functional statistical framework. In the various examples presented in the three next sections we will give explicit evaluations of the concentration function φ_χ, and we will discuss how the general results presented all through the book are behaving. The main point of the statistical methodology described in this book is to allow for a wide scope of possible topological structures, since the space E is only assumed to be of semi-metric type, and we will see (specifically in Section 13.3) all the advantages of this general modelling compared with what one would get with more popular (but too much restrictive) Banach or Hilbert type structures.

13.3 Some Special Infinite Dimensional Processes

To highlight our purpose, we will focus on two special classes of functional variables as defined below in Definitions 13.1 and 13.4. We will see in Section 13.3.1 that all the methology developed earlier in this book applies for fractal-type processes. Then, we will look in Section 13.3.2 to an other class which is known to contain main continuous time processes and for which the overdispersed form of the small ball probability function does not allow us to expect good results (at least directly) from our nonparametric functional methodology. Finally in Section 13.3.3 we will come back to the role of the semi-metric and to its links with the small probability function. In particular we will show how we can construct in any situation a new semi-metric for which the process is always of fractal-type and this allows us to expect good results of the functional nonparametric method (depending, of course, on this topological structural change), in any case (including the overdispersed ones discussed in Section 13.3.2).

13.3.1 Fractal-type Processes

Recall that χ is fixed and note that in the following definition the constants τ and C may be depending on χ. From a chronological point of view, the notion of fractal dimension is closely related with some other ones widely used in the physical sciences. The following definition was previously introduced in [FV00] for functional nonparametric purposes (see also [BL01] for a finite dimensional interest of this notion).

Definition 13.1. *The variable $\mathcal{X}$ is said to be of fractal order τ, with respect to the semi-metric d, if there exists some finite constant $C > 0$ such that the associated concentration function φ_χ is of the form*

$$\varphi_\chi(\epsilon) \sim C\epsilon^\tau \text{ as } \epsilon \to 0.$$

The reader will find in [P93] a general presentation of various other notions of dimension useful in physics, and the gap between these notions and Definition 13.1 below is described in Section 4 of [FV00]. This notion is of particular interest since for fractal processes all the functional nonparametric estimates discussed before in this book can be shown to reach the same kind of rates of convergence as for finite dimensional setting. This is formalized in Proposition 13.2 below. To avoid tedious repetitions we just show how the results obtained with i.i.d. variables on functional kernel regression, functional kernel conditional mode and functional kernel conditional quantile behave in the fractal case. The reader will see easily that the same kind of results could be obtained from Proposition 6.19 in conditional c.d.f. estimation, or from Proposition 6.20 in conditional density estimation, or from Theorem 8.2 in curves discrimination problems. Similarly, one could do the same kind of things by using the results obtained in Part IV of this book for α-mixing situations.

Proposition 13.2. *Assume that $\mathcal{X}$ is of fractal order τ.*

i) *Regression. Under the conditions of Theorem 6.11, the functional kernel regression estimate can reach the rate of convergence:*

$$\widehat{r}(\chi) - r(\chi) \;=\; O_{a.co.}\left(\left(\frac{\log n}{n}\right)^{\frac{\beta}{2\beta+\tau}}\right).$$

ii) *Mode. Under the conditions of Theorem 6.16, the functional kernel mode estimate can reach the rate of convergence:*

$$\widehat{\theta}(\chi) - \theta(\chi) \;=\; O_{a.co.}\left(\left(\frac{\log n}{n}\right)^{\frac{\beta}{2j\beta+\tau+1}}\right).$$

iii) *Quantile. Under the conditions of Theorem 6.18, the functional kernel quantile estimate can reach the rate of convergence:*

$$\widehat{t}_\alpha(\chi) - t_\alpha(\chi) \;=\; O_{a.co.}\left(\left(\frac{\log n}{n}\right)^{\frac{\beta}{2j\beta+\tau}}\right).$$

Proof. It suffices to combine the result of each among the mentioned theorems together with the expression of the small ball probability function $\varphi_\chi(.)$ and:

i) for regression estimation, choose the bandwidth in the following way:

$$h_K \sim C\left(\frac{\log n}{n}\right)^{\frac{1}{2\beta+\tau}};$$

ii) for mode estimation, choose the bandwidths in the following way:

$$h_K \sim h_H \sim C \left(\frac{\log n}{n} \right)^{\frac{1}{2j\beta + \tau + 1}} ;$$

iii) for quantile estimation, choose the bandwidths in the following way:

$$h_K \sim h_H \sim C \left(\frac{\log n}{n} \right)^{\frac{1}{2j\beta + \tau}} .$$

This is enough to complete this proof. $\square$

It is worth noting that the rates of convergence presented in Proposition 13.2 are similar to those already existing in finite dimensional setting (see details in Section 13.5 below). This appealing feature will be strenghtened in Section 13.3.3 where it will be seen that a good choice of the semi-metric d can always insure that the variable $\mathcal{X}$ is of fractal-type.

Remark 13.3. Note that the condition (4.6), which plays a key role in applying our methodology with continuous kernels of type II, is obviously satisfied for fractal-type processes.

13.3.2 Exponential-type Processes

In the recent probabilistic literature, many works have been devoted to the statement of asymptotic evaluation of small ball probabilities for various famous continuous time stochastic processes. It turns out that, staying with standard metric spaces, many usual processes are of the following exponential-type. Recall that χ is fixed and note that in the following definition the constants τ_1, τ_2, and C may be depending on χ.

Definition 13.4. *The variable $\mathcal{X}$ is said to be of exponential-type with orders (τ_1, τ_2), with respect to the semi-metric d, if there exists some finite constant $C > 0$ such that the associated concentration function φ_χ is of the form*

$$\varphi_\chi(\epsilon) \sim Ce^{-\frac{1}{\epsilon^{\tau_1}} \log(\frac{1}{\epsilon})^{\tau_2}} \text{ as } \epsilon \to 0.$$

To highlight the interest of this definition, let us look at two classes of examples both being concerned with the space of the real-valued continuous functions $\mathcal{C}([0,1])$ endowed with the metric d associated with the supremum norm:

$$\forall x \in \mathcal{C}([0,1]), \ ||x||_{sup} = \sup_{t \in [0,1]} |x(t)|.$$

It is out of the scope of this book to provide details about these examples. Our main wish is to emphasize their exponential small ball probabilities forms and on their impacts in the statistical framework. The first class of examples concern diffusion processes. The reader will find complementary discussions in [B78], [LS72], [B99], [D02] and [FLV05]. According to the results in [B99], the usual Ornstein-Uhlenbeck diffusion process satisfies the Definition 13.4 with orders $\tau_1 = 2$ and $\tau_2 = 0$. This author shows firstly this result when $\chi = 0$ (see page 187 in [B99]), and extension to any other χ in the associated Cameron-Martin space is available (see Chapter 2 in [B99]). The same kind of result can be extended to many diffusion processes being absolutely continuous with respect to the Wiener measure (see [LS72] for characterizations of such absolutely continuous diffusion processes). The second class of examples concern Gaussian processes, and the reader will find more complete informations in [B99], [LS01], and [FLV05]. Let us just mention that the exponential form as given in Definition 13.4 has been obtained by [LS01] for fractional Brownian motion(with orders τ_j depending on the parameter of the Brownian motion). This standard result was extended for processes in $\mathcal{C}([0,1]^q)$, for some $q \in \mathbb{N}^*$ (see Corollary 4.10.7 in [B99] and Theorem 4.6 in [LS01]). It concerns for instance Lévy fractional motions and fractional Brownian sheets (see [LS01] and [BL02]) and fractional Ornstein-Uhlenbeck processes (see Example 4.10.9 in [B99]). To conlude this short survey of examples of exponential-type processes, let us just say that it is a current field of investigation in modern probability theory, and in the last few years many other processes have been concerned with the statement of their small ball probability functions. All of them have been shown to be of the exponential-type described in Definition 13.4. A nice overview of the recent infatuation for this kind of problem in the probability literature can be found in [NN04], [GLP04], [GHLT04], [GHT03], [S03], [D03], [CL03], [DFMS03], [BL02] and references therein.

It should be noted that this common exponential feature for the small ball probabilities of time continuous processes is not linked with the choice of the supremum norm introduced before. Indeed, the same kind of results can be obtained under different topological structures. For instance, by changing the supremum norm into the L_s-norm defined for some $s > o$ by

$$||x||_{L_s} = \left(\sup_t |x(t)|^s \right)^{\frac{1}{s}},$$

similar results are derived by [NN04], [GHLT04], [GHT03]. The case of Hölder norms H_α defined for some $\alpha > 0$

$$||x||_\alpha = \sup_{t \neq t'} \frac{|x(t) - x(t')|}{|t - t'|^\alpha},$$

and more generally the case of f-norm defined from some real-valued function f by

$$||x||_f = \sup_{t \neq t'} \frac{|x(t) - x(t')|}{f(|t - t'|)},$$

are investigated in [B99] (see Theorem 4.10.6). As far as we know, there is no result of this kind in semi-metric spaces.

The functional nonparametric methodology works without specifying any form for the small ball probability function $\varphi_\chi(.)$, and so it concerns the case of exponential-type processes such as those described earlier. The next proposition shows how the rates of convergence of the functional kernel estimates behave for such processes. As in Proposition 13.2 for fractal processes, we just discuss the results for i.i.d. variables and for kernel regression, conditional mode and conditional quantile estimates. The reader will see easily that the same kind of results could be obtained from Proposition 6.19 in conditional c.d.f. estimation, from Proposition 6.20 in conditional density estimation, from Theorem 8.2 in discrimination problems, or for α-mixing variables from the results obtained in Part IV of this book.

Proposition 13.5. *Assume that $\mathcal{X}$ is of exponential-type.*

i) *Regression. Under the conditions of Theorem 6.11, the functional kernel regression estimate can reach a rate of convergence of the form:*

$$\widehat{r}(\chi) - r(\chi) \; = \; O_{a.co.}\left((\log n)^{-v_1}\right), \; \textit{for some } v_1 > 0.$$

ii) *Mode. Under the conditions of Theorem 6.16, the functional kernel mode estimate can reach a rate of convergence of the form:*

$$\widehat{\theta}(\chi) - \theta(\chi) = O_{a.co.}\left((\log n)^{-v_2}\right), \; \textit{for some } v_2 > 0.$$

iii) *Quantile. Under the conditions of Theorem 6.18, the functional kernel quantile estimate can reach a rate of convergence of the form:*

$$\widehat{t}_\alpha(\chi) - t_\alpha(\chi) \; = \; O_{a.co.}\left((\log n)^{-v_3}\right), \; \textit{for some } v_3 > 0.$$

Proof. It suffices to combine the result of each among the mentioned theorems together with the expression of the small ball probability function $\varphi_\chi(.)$ and to choose bandwidths of the form:

$$h_K \sim h_H \sim C \left(\log n\right)^u, \; \textit{for some } u < 0.$$

Note that the exponent u is different for the three different assertions of the theorem. $\square$

Even if the optimality of the rates of convergence obtained in Proposition 13.5 is still to be proved, as pointed out by [D02] in a related framework, one may reasonably hope this to be true. This guess is based on the fact that for p-dimensional problems and for β-Lipschitz models the optimal rates of convergence are known to be of order $(\log n/n)^{\beta/(2\beta+p)}$ (see [S82]). This supports the idea that in infinite dimensional framework the rate of convergence cannot be a power of n. This is known as the *curse of the infinite dimension* (see

[FV03b] for more extensive discussion). This question could be formulated as follows:

Open question 13 *For nonparametric problems involving exponential-type processes, are the optimal rates of convergence of the form* $(\log n)^u$ *for some $u < 0$?*

Of course, this open problem is of interest for deeper understanding of the probabilistic phenomena. Anyway, one can immediately draw some interesting statistical conclusions. In a first attempt, Proposition 13.5 could look quite disappointing because rates of convergence as powers of $\log n$ are not satisfactory from a statistical point of view. Indeed, we will see in the next section that looking more deeply at the statistical significance of the results, one could quickly change their mind.

13.3.3 Links with Semi-metric Choice

Indeed, the small ball probabilities functions are directly linked with the concentration properties of the functional variable $\mathcal{X}$. Similar to what happens in p-dimensional problems with the curse of dimensionality, the poor rates of convergence derived in Proposition 13.5 can be explained by an overdispersion phenomenon of exponential-type processes linked with the dramatically fast decaying of the ball probability function φ_χ around 0. On the other hand, the more appealing rates stated in Proposition 13.2 for fractal processes are linked with the slower decaying of the ball probability function φ_χ around 0. These facts agree completely with the empirical ideas developed in Section 3.3.

Because the topological structure controls the concentration properties, the natural answer that comes to mind is to try to change the topology. At first, looking at the current probabilistic information on exponential-type processes discussed before, the small ball probability function is always of exponential form, as well for the topology associated with the supremum norm as for those associated with L_s, H_α or f norms. This highlights powerfully the idea that, to be efficient in terms of higher concentration of the variable $\mathcal{X}$, a new topological structure should not be driven by the standard metric procedures. This is a strong theoretical motivation for developing (as far as possible) the functional nonparametric methodology in semi-metric spaces. Once again, this agrees strongly with the empirical arguments developed in Chapter 3. To show that a topological structure based on a suitable semi-metric is effectively efficient for increasing the concentration properties of the variable $\mathcal{X}$, we will now indicate a general procedure allowing us to construct a semi-metric for which the process $\mathcal{X}$ is necessarily of fractal-type (according to this new topology).

Lemma 13.6. *Let $\mathcal{H}$ be a separable Hilbert space with inner product $< .,.>$ and let $\{e_j, j = 1, \ldots \infty\}$ an orthonormal basis. Let $k \in \mathbb{N}^*$ be fixed. Let $\chi = \sum_{j=1}^{\infty} x^j e_j$ be a fixed element in $\mathcal{H}$.*

i) The function defined by
$$\forall (\chi', \chi'') \in \mathcal{H} \times \mathcal{H}, \ d_k(\chi', \chi'') = \sqrt{\sum_{j=1}^{k} < \chi' - \chi'', e_j >^2},$$
is a semi-metric on the space $\mathcal{H}$.

ii) Let $\mathcal{X} = \sum_{j=1}^{\infty} X^j e_j$ be a squared integrable random element of $\mathcal{H}$. If the random variable $X = (X^1, \ldots, X^k)$ is absolutely continuous with respect to the Lebesgues measure on $\mathbb{R}^k$ with a density function f being continuous at point $\mathbf{x} = (x^1, \ldots x^k)$ and such that $f(\mathbf{x}) > 0$, then the process $\mathcal{X}$ is of fractal order k with respect to the semi-metric d_k, in the sense of Definition 13.1.

Proof. i). For any $(\chi', \chi'', \chi''') \in \mathcal{H} \times \mathcal{H} \times \mathcal{H}$ let us denote by $(\boldsymbol{X}', \boldsymbol{X}'', \boldsymbol{X}''')$ the associated k-dimensional real vectors $\boldsymbol{X}' = (< \chi', e_1 >, \ldots < \chi', e_k >)$, $\boldsymbol{X}'' = (< \chi'', e_1 >, \ldots < \chi'', e_k >)$ and $\boldsymbol{X}''' = (< \chi''', e_1 >, \ldots < \chi''', e_k >)$. If we denote by d_{eucl} the euclidian metric on $\mathbb{R}^k$, we can write:

$$d_k(\chi', \chi'') = d_{eucl}(\boldsymbol{X}', \boldsymbol{X}'').$$

So, using the properties of the metric d_{eucl}, we have both of the following results:

$$\begin{aligned}
\chi' = \chi'' &\Rightarrow \boldsymbol{X}' = \boldsymbol{X}'' \\
&\Rightarrow d_{eucl}(\boldsymbol{X}', \boldsymbol{X}'') = 0 \\
&\Rightarrow d_k(\chi', \chi'') = 0,
\end{aligned} \tag{13.1}$$

and

$$\begin{aligned}
d_k(\chi', \chi'') &= d_{eucl}(\boldsymbol{X}', \boldsymbol{X}'') \\
&\leq d_{eucl}(\boldsymbol{X}', \boldsymbol{X}''') + d_{eucl}(\boldsymbol{X}''', \boldsymbol{X}'') \\
&= d_k(\chi', \chi''') + d_k(\chi''', \chi'').
\end{aligned} \tag{13.2}$$

According to Definition 3.2, these are both conditions needed to insure that d_k is a semi-metric.

ii). Let us compute the small ball probability function associated with the process $\mathcal{X}$ and the semi-metric d_k. For $\epsilon > 0$ we have:

$$\varphi_\chi(\epsilon) = P\left(d_k(\boldsymbol{\mathcal{X}}, \chi) \le \epsilon\right)$$

$$= P\left(\sqrt{\sum_{j=1}^{k} <\boldsymbol{\mathcal{X}} - \chi, e_j>^2} \le \epsilon\right)$$

$$= P\left(\sqrt{\sum_{j=1}^{k} (X^j - x^j)^2} \le \epsilon\right)$$

$$= P\left(\|\boldsymbol{X} - \boldsymbol{x}\| \le \epsilon\right),$$

where $\|.\|$ is the usual euclidian norm on $\mathbb{R}^k$. By using now the continuity condition on f, and by using the notation $V(k)$ for the volume of the unit ball in $\mathbb{R}^k$, we arrive at:

$$\varphi_\chi(\epsilon) = \int_{B(\boldsymbol{x},\epsilon)} f(\boldsymbol{t}) d\boldsymbol{t}$$

$$= \int_{B(\boldsymbol{x},\epsilon)} d\boldsymbol{t} \left(f(\boldsymbol{x}) + O(\epsilon)\right)$$

$$= \epsilon^k V(k) f(\boldsymbol{x}) + o(\epsilon^k).$$

$$= \epsilon^k \frac{\pi^{\frac{k}{2}}}{\frac{k}{2}\Gamma(\frac{k}{2})} f(x) + o(\epsilon^k).$$

This enough to show that, for the semi-metric d_k, the fractal property introduced in Definition 13.1 holds.□

Note that d_k is not a metric since the reverse of the property (13.1) is obviously false. This result is particularly appealing for answering the semi-metric choice question,because it shows that in any case we can construct some semi-metric for which the process could be considered of fractal-type, and so for which the nonparametric methodology exhibits the rates of convergence specified in Proposition 13.2. In particular, even for the overdispersed exponential processes, the unappealing rates obtained in Proposition 13.5 can be easily surpassed by the topological structure induced by such a new semi-metric.

The semi-metrics described in Lemma 13.6 are usually known as *projections type semi-metrics*. They can be constructed in various different ways according to the space $\mathcal{H}$ and to its selected orthonormal basis. For instance it concerns Fourier basis, as various wavelet bases, as well as the Functional PCA projection semi-metric described in Section 3.4.1 and which has been applied succesfully to the phoneme data in Section 8.4.

13.4 Back to the One-dimensional Setting

Even if the scope of this book is to develop modelling and statistical methods for infinite dimensional problems, the methodology concerns variables taking values in any abstract semi-metric space. So, all the results presented before can be applied in the one-dimensional setting when $E = \mathbb{R}$. Let us for instance choose d to be the usual euclidian metric on the real line. In the following, according to the general notations presented in Chapter 1.4, we will use the notation X (in place of $\mathcal{X}$) for the real random variable and we will use x (in place of χ) to denote a fixed deterministic element of $\mathbb{R}$.

In a first attempt, let us look at the situation when X is absolutely continuous with respect to Lebesgue measure, and with a density f satisfying:

$$f \text{ is continuous and } f(x) > 0. \tag{13.3}$$

We will see that such a situation corresponds to a fractal process of order 1 in the sense of Definition 13.1.

Lemma 13.7. *If X satisfies (13.3) then its small ball probability function is such that for some $C > 0$:*

$$\varphi_x(\epsilon) = C\epsilon + o(\epsilon).$$

Proof. The small ball probability function is defined as

$$\varphi_x(\epsilon) = P\left(|X - x| \le \epsilon\right) = \int_{x-\epsilon}^{x+\epsilon} f(t)dt,$$

and the continuity of f allows us to conclude directly by taking $C = 2f(x)$.

From this result, one can find again many results already stated in the classical one-dimensional nonparametric literature. For instance, by combining this lemma with Proposition 13.2 we get the following results.

> **Proposition 13.8.** *Assume that X satisfies (13.3).*
>
> *i) Regression. Under the conditions of Theorem 6.11, the functional kernel regression estimate can reach the rate of convergence:*
>
> $$\widehat{r}(x) - r(x) = O_{a.co.}\left(\left(\frac{\log n}{n}\right)^{\frac{\beta}{2\beta+1}}\right).$$
>
> *ii) Mode. Under the conditions of Theorem 6.16, the functional kernel mode estimate can reach the rate of convergence:*
>
> $$\widehat{\theta}(x) - \theta(x) = O_{a.co.}\left(\left(\frac{\log n}{n}\right)^{\frac{\beta}{2j\beta+2}}\right).$$
>
> *iii) Quantile. Under the conditions of Theorem 6.18, the functional kernel quantile estimate can reach the rate of convergence:*
>
> $$\widehat{t}_\alpha(x) - t_\alpha(x) = O_{a.co.}\left(\left(\frac{\log n}{n}\right)^{\frac{\beta}{2j\beta+1}}\right).$$

Proof. It is a direct consequence of Proposition 13.2 and Lemma 13.7.□

Note for instance that, for the nonparametric kernel regression estimate, the usual rate of convergence of order $(\log n/n)^{\beta/(2\beta+1)}$ has been shown to be optimal by [S82]. Similar conclusions can be drawn for the other problems studied in this book, for density or c.d.f. estimation, as well as for classification problems and for α-mixing variables, just by plugging the expression of the small ball probability function given by Lemma 13.7 into each of the rates of convergence stated earlier in the book.

The last (but not least) point is to see that the application of the general methodology to one-dimensional setting may be drawn even for r.r.v. variables which do not satisfy the conditions of Lemma 13.7. Curiously, almost all the dramatically abundant literature in the finite-dimensional nonparametric literature works only with variables like those in Lemma 13.7. In both lemmas below we will describe two situations which are not concerned with the standard one-dimensional literature but for which the general functional methodology will apply directly. Let us first look at the case where the distribution of X is not absolutely continuous but such that

$$P(X = x) = \delta > 0. \tag{13.4}$$

Lemma 13.9. *If X satisfies (13.4) then its small ball probability function is such that there exists some $C > 0$ such that for any $\epsilon > 0$:*

$$\varphi_x(\epsilon) \geq C.$$

Proof. This result is true with $C = \delta$. It suffices to write:

$$\varphi_x(\epsilon) = P\left(|X - x| \leq \epsilon\right)$$
$$\geq P(X = x) = \delta > 0. \quad \square$$

For many of the problems studied in this book, including conditional quantile, mode, density or c.d.f. estimation as well as discrimination problems, we can get from this lemma rates of convergence for kernel estimates of discontinuous real variable X whose c.d.f. has jumps. Some of these results are reported in the next proposition.

Proposition 13.10. *Assume that X satisfies (13.4).*

i) Regression. Under the conditions of Theorem 6.11, the functional kernel regression estimate can reach the rate of convergence:

$$\widehat{r}(x) - r(x) = O\left(h^{\beta}\right) + O_{a.co.}\left(\sqrt{\frac{\log n}{n}}\right).$$

ii) Mode. Under the conditions of Theorem 6.16, the functional kernel mode estimate can reach the rate of convergence:

$$\widehat{\theta}(x) - \theta(x) = O\left(\left(h^{\beta} + g^{\beta}\right)^{\frac{1}{j}}\right) + O_{a.co.}\left(\left(\frac{\log n}{ng}\right)^{\frac{1}{2j}}\right).$$

iii) Quantile. Under the conditions of Theorem 6.18, the functional kernel quantile estimate can reach the rate of convergence:

$$\widehat{t}_{\alpha}(x) - t_{\alpha}(x) = O\left(\left(h^{\beta} + g^{\beta}\right)^{\frac{1}{j}}\right) + O_{a.co.}\left(\left(\frac{\log n}{n}\right)^{\frac{1}{2j}}\right).$$

Proof. Each assertion follows by plugging the small ball probability expression given in Lemma 13.9 into the corresponding theorem.$\square$

Of course, any other result presented before in the book could be re-expressed similarly for a real random variable with jumps. Indeed, there are only two kinds of results which cannot be applied to discontinuous variables. The first one concerns unsupervised classification, and since the method described in Chapter 9 is completely based on the existence of some density function there is no wish (nor interest) to expect such application to be possible. The second exception concerns the dependent situations studied throughout Part IV of this book but here the reason is purely of technical order and comes from the additional conditions (see for instance (11.12)) which are excluding discontinuous situations. One could reasonably hope to improve the way we derived our calculus in Part IV in order to allow for variables of the kind described in Lemma 13.9. Anyway, excepting these two special caes, a direct use of the functional methodology to the specific one-dimensional framework allows us to extend many classical results to variables X being not necessarily absolutely continuous with respect to Lebesgue measure.

Let us now present another kind of real random variable, which is not covered in the usual one-dimensional nonparametric literature, but for which the functional methodology applies directly. This case concerns variables X having a distribution function F such that:

$$\exists \tau > 0, \exists 0 < g_x < \infty, \ F(x + \epsilon) - F(x - \epsilon) = g_x \epsilon^\tau + o(\epsilon^\tau). \qquad (13.5)$$

Lemma 13.11. *If X satisfies (13.5) then its small ball probability function is such that:*

$$\varphi_x(\epsilon) \ = \ C\epsilon^\tau + o(\epsilon^\tau).$$

Proof. This result is obvious since:

$$\varphi_x(\epsilon) = P\left(|X - x| \le \epsilon\right) = F(x + \epsilon) - F(x - \epsilon). \square$$

Once again note that any of the various results of this book can be applied directly, just by plugging in the expression of the small ball probability function given by Lemma 13.11. The next proposition states some of these results. These results are particularly interesting because the usual one-dimensional statistical literature is only concerned with the special case when $\tau = 1$, which corresponds exactly to the standard situation described in Lemma 13.7. For instance, the case when $\tau < 1$ allows for variables X having a distribution function which is not differentiable at point x while, conversely, the case when $\tau > 1$ allows for variables X having a density function which is vanishing at point x.

Proposition 13.12. *Assume that X statisfies (13.5).*

i) *Regression. Under the conditions of Theorem 6.11, the functional kernel regression estimate can reach the rate of convergence:*

$$\widehat{r}(x) - r(x) = O_{a.co.} \left(\left(\frac{\log n}{n} \right)^{\frac{\beta}{2\beta+\tau}} \right).$$

ii) *Mode. Under the conditions of Theorem 6.16, the functional kernel mode estimate can reach the rate of convergence:*

$$\widehat{\theta}(x) - \theta(x) = O_{a.co.} \left(\left(\frac{\log n}{n} \right)^{\frac{\beta}{2j\beta+\tau+1}} \right).$$

iii) *Quantile. Under the conditions of Theorem 6.18, the functional kernel quantile estimate can reach the rate of convergence:*

$$\widehat{t}_\alpha(x) - t_\alpha(x) = O_{a.co.} \left(\left(\frac{\log n}{n} \right)^{\frac{\beta}{2j\beta+\tau}} \right).$$

Proof. It is a direct consequence of Proposition 13.2 and Lemma 13.11. $\square$

13.5 Back to the Multi- (but Finite) -Dimensional Setting

Let us look now at how the results presented before can be applied in the multi-dimensional nonparametric setting when $E = \mathbb{R}^p$. In the following, according to the general notations presented in Chapter 1.4, we will use the notation X (in place of $\mathcal{X}$) for the multivariate random variable and we will use x (in place of χ) to denote a fixed deterministic element of $\mathbb{R}^p$.

In a first attempt, let us look at the situation when d is the usual euclidian metric on $E = \mathbb{R}^p$ and when X is absolutely continuous with respect to Lebesgue measure with density f satisfying:

$$f \text{ is continuous and such that } f(x) > 0. \tag{13.6}$$

We will see that such a situation corresponds to a fractal process of order p in the sense of Definition 13.1.

Lemma 13.13. *If X satisfies (13.6) then its small ball probability function is such that for some $C > 0$:*

$$\varphi_x(\epsilon) = C\epsilon^p + o(\epsilon^p).$$

Proof. The proof is easily obtained with

$$C = \frac{\pi^{\frac{p}{2}}}{\frac{p}{2}\Gamma(\frac{p}{2})},$$

according to the following steps:

$$\begin{aligned}
\varphi_{\boldsymbol{x}}(\epsilon) &= \int_{B(\boldsymbol{x},\epsilon)} f(\boldsymbol{t})d\boldsymbol{t} \\
&= \int_{B(\boldsymbol{x},\epsilon)} d\boldsymbol{t} \left(f(\boldsymbol{x}) + O(\epsilon) \right) \\
&= \epsilon^p V(p) f(\boldsymbol{x}) + o(\epsilon^p). \\
&= \epsilon^p \frac{\pi^{\frac{p}{2}}}{\frac{p}{2}\Gamma(\frac{p}{2})} f(x) + o(\epsilon^k).
\end{aligned}$$

Here, we denoted by $V(p)$ the volume of the unit ball in $\mathbb{R}^p$ (for the topology associated with the euclidian metric).□

From this result, as we did in Section 13.4 when $p = 1$, one can find again many results of the classical p-dimensional nonparametric literature. The next proposition will summarize some of these possible results.

Proposition 13.14. *Assume that* $\boldsymbol{X}$ *satisfies (13.6).*

i) *Regression. Under the conditions of Theorem 6.11, the functional kernel regression estimate can reach the rate of convergence:*

$$\widehat{r}(\boldsymbol{x}) - r(\boldsymbol{x}) = O_{a.co.}\left(\left(\frac{\log n}{n} \right)^{\frac{\beta}{2\beta+p}} \right).$$

ii) *Mode. Under the conditions of Theorem 6.16, the functional kernel mode estimate can reach the rate of convergence:*

$$\widehat{\theta}(\boldsymbol{x}) - \theta(\boldsymbol{x}) = O_{a.co.}\left(\left(\frac{\log n}{n} \right)^{\frac{\beta}{2j\beta+p+1}} \right).$$

iii) *Quantile. Under the conditions of Theorem 6.18, the functional kernel quantile estimate can reach the rate of convergence:*

$$\widehat{t}_\alpha(\boldsymbol{x}) - t_\alpha(\boldsymbol{x}) = O_{a.co.}\left(\left(\frac{\log n}{n} \right)^{\frac{\beta}{2j\beta+p}} \right).$$

Proof. It is a direct consequence of Proposition 13.2 and Lemma 13.13.□

Note for instance that, for the nonparametric kernel regression estimate, the usual rate of convergence of order $(\log n/n)^{\beta/(2\beta+p)}$ has been shown to be optimal by [S82]. Similar conclusions can be drawn for the other problems studied in this book, for kernel conditional density or c.d.f. estimation, for discrimination or classification problems, as well as for α-mixing variables, just by plugging the expression of the small ball probability function given by Lemma 13.13 into each of the rates of convergence stated earlier in the book. Of course, as in Section 13.4 when $p = 1$, we could derive results in less standard situations (for instance like those described in Lemma 13.11) and show that the functional methodology would allow the extension of the usual nonparametric literature to less restrictive conditions on the probability distribution of X. We will not discuss this in detail here, both because the discussion would be a tedious repetition of Section 13.4 and also because there is much more to win in the p-dimensional framework than a simple reduction of the hypothesis.

Indeed, the general functional methodology allows us to attack the curse of dimensionality from a new point of view. This is what we will discuss now through the following result:

Lemma 13.15. *Let $\{e_j,\, j = 1, \ldots p\}$ be an orthonormal basis of $\mathbb{R}^p$. Let $k \in \{1, \ldots p-1\}$ be fixed, and write $x = \sum_{j=1}^{p} x^j e_j$.*

i)   The function defined for $(\mathbf{y}, \mathbf{z}) = ((y^1, \ldots y^p), (z^1, \ldots z^p)) \in \mathbb{R}^p \times \mathbb{R}^p$ by
$$d_k(\mathbf{y}, \mathbf{z}) = \sqrt{\sum_{j=1}^{k}(y^j - z^j)^2},$$
is a semi-metric on $\mathbb{R}^p$.

ii) Let $X = \sum_{j=1}^{p} X^j e_j$ be a squared integrable random variable in $\mathbb{R}^p$, such that $(X^1, \ldots, X^k)$ is absolutely continuous with respect to the Lebesgues measure on $\mathbb{R}^k$ with a density function f being continuous and such that $f(x^1, \ldots x^k) > 0$. Then, the process x is of fractal order k with respect to the semi-metric d_k.

Proof. This result is just a special case of Lemma 13.6.$\square$

This result is particularly appealing for the semi-metric choice problem because it shows that we can construct some semi-metric for which the process could be considered of fractal-type with order $k < p$. We will see that, if we use such a semi-metric, the nonparametric methodology exhibits rates of convergence faster than with the usual euclidian metric. Of course, this does not contradict the results in [S82] about the optimality of the rate $(\log n/n)^{\beta/(2\beta+p)}$, since changing the semi-metric has changed the model (because the smoothness Lipschitz model on the regression function r does not have the same meaning from one topology to the other). The next proposition will summarize some results of this type that can be obtained by combining our general functional methodology with Lemma 13.15.

Proposition 13.16. *If* X *satisfies the condition ii) of Lemma 13.15:*

i) Regression. Under the conditions of Theorem 6.11, the functional kernel regression estimate can reach the rate of convergence:

$$\widehat{r}(x) - r(x) \;=\; O_{a.co.}\left(\left(\frac{\log n}{n}\right)^{\frac{\beta}{2\beta+k}}\right).$$

ii) Mode. Under the conditions of Theorem 6.16, the functional kernel mode estimate can reach the rate of convergence:

$$\widehat{\theta}(x) - \theta(x) \;=\; O_{a.co.}\left(\left(\frac{\log n}{n}\right)^{\frac{\beta}{2j\beta+k+1}}\right).$$

iii) Quantile. Under the conditions of Theorem 6.18, the functional kernel quantile estimate can reach the rate of convergence:

$$\widehat{t}_\alpha(x) - t_\alpha(x) \;=\; O_{a.co.}\left(\left(\frac{\log n}{n}\right)^{\frac{\beta}{2j\beta+k}}\right).$$

Proof. It is a direct consequence of Proposition 13.2 and Lemma 13.15.□

For instance, staying only in regression setting to make discussion shorter and clearer, the nonparametric estimate has a rate of convergence of order $(\log n/n)^{\beta/(2\beta+k)}$ which is evidently better than the order $(\log n/n)^{\beta/(2\beta+p)}$ obtained before with the euclidian metric. Note that in the extreme case, with $k = 1$, we can reach a rate of convergence which is the same as in the univariate case, and which is therefore independent of the dimensionality p of the problem. So, there is real evidence for saying that such an approach is a good candidate for constructing new models for dimension reduction in multivariate nonparametric framework.

Finally, to fix the ideas let us mention that a typical example of such a semi-metric would be obtained by using as basis $\{e_j, j = 1, \ldots p\}$ the one obtained by some standard multivariate Principal Component Analysis. Of course, many other choices are possible. Each choice of the basis will lead to a different projection semi-metric and then to a new model. In other words, this supports the idea that many new models for dimension reduction can be constructed just by changing the topological structure (that is, by introducing some projection type semi-metric such as is defined in Lemma 13.15). It is evidently out of the scope of this book to enter more deeply into this discussion, but we share the point of view that some interesting further advances could be obtained this way for reducing dimensional effects in multi- (but finite) -dimensional nonparametric problems.

13.6 The Semi-metric: a Crucial Parameter

Without any doubt, the semi-metric appears to be a key parameter for insuring the good behaviour of any nonparametric statistical method on functional data. The empirical ideas developed in Chapter 3 gave forewarning of this point, and this is evidently confirmed with all the theoretical developments presented in this chapter. Because of Proposition 13.6, we have now at hand for any functional variable $\mathcal{X}$ a way to construct one (or more than one) semi-metric for which the small ball probability function is of fractal-type (see Definition 13.1), and we know (see Proposition 13.2) that in such a case the functional nonparametric methods have rates of convergence of the same kind as in finite-dimensional problems. This approach will be interesting when $\mathcal{X}$ is of an exponential form (see Definition 13.4) or of some related ones, since in this case the usual metric modelling is not efficient (see Proposition 13.5).

At this stage, theoretical advances for choosing the semi-metric in practical situations have not yet been developed. Because of the difficulty of the problem (linked in particular with the wide set of possible semi-metrics), we support the idea that this choice has absolutely to be driven by taking into account some non-statistical information about the process $\mathcal{X}$. This general guideline was followed all through the case studies presented before. For instance, going back to the spectrometric data presented in Section 2.1, their quite smooth feature suggests a semi-metric based on the derivatives of the curves. This choice gave nice results in prediction (see Chapter 7) as well as in classification problems (see Chapter 8). The quite different unsmooth shape of the phoneme data presented in Section 2.2 evidently supported the idea that such a semi-metric based on derivatives of the curves was not competitive, and has oriented through the use of some projection-type semi-metric. This gave nice results on these data (see Chapters 8 and 9). In the same spirit, but for other reasons linked with the few numbers of discretized points, the electrical consumption data presented in Section 2.3 have been treated by a projection approach and the results were pretty appealing (see Chapter 12).

Finally, one would give the following practical general advice for users. For smooth curves, the semi-metric based on higher order derivatives could be a good one, while for unsmooth or for sparsely observed curves an approach based on a projection type semi-metric will be more efficient. In some specific problems one could have additional information that can be incorporated into the construction of the semi-metric. This was the case in discrimination, for which the knowledge of the group membership of each curve had led to a new semi-metric based on PLS ideas (see Chapter 8). Of course, depending on the kind of data to be treated and on the statistical problem investigated, the user can construct his or her own semi-metric. Such a new semi-metric could be easily integrated in the general software which goes with this book[1].

---

[1] http://www.lsp.ups-tlse.fr/staph/npfda

14

Some Perspectives

The large quantity of recent international publications on statistical methods and models for functional data emphasizes the great interest in this field shared by statisticians and other users. Therefore this monograph has been written in order to present the main ideas by discussing both their practical impacts and their theoretical senses. This is particularly the case with the "semi-metrics", "local functional weighting" and "nonparametric modelling" which play major roles in the new nonparametric methodology for functional data developed all through this book. The feedback practice/theory strengthens the bridge between practitioners and theoreticians, which is at the heart of any applied mathematical activities, and it is a key point for ensuring development of the knowledge on functional statistics. As a good illustration of this double objective, this book is strongly linked with the companion website (http://www.lsp.ups-tlse.fr/staph/npfda) which proposes easy implementation of the functional nonparametric methods.

Because of the novelty of such a statistical area, it is clear that one can expect many further advances in the next few years. Beyond practical aspects (increase of the R/S+ current library and/or translation into other languages) and technical theoretical open questions pointed out throughout the book, a great challenge for the future will be to extend the functional statistical methods to functional data more complex than curves (surfaces, arrays of functional data, images, . . .). The mathematical background described in this book allows for nonparametric analyses of such objects whereas practical aspects remain to be developed (in particular the building of semi-metrics adapted to this kind of data will certainly play a key role in the future). Of course, other new questions will emerge from such intricate functional data (in particular, spatial notions are often involved for analyzing images and a great field of new investigations will concern spatial functional data).

Finally, we hope that this book will contribute to the dissemination of recent knowledge on nonparametric functional statistics, will motivate further advances in this field and will popularize these new methods in many scientific communities having to deal with real functional datasets.

Appendix

Some Probabilistic Tools

In order to give the complete mathematical background, we decided to present briefly some probabilistic tools. Some of these tools have been formulated in new ways to make them easily applicable for the functional nonparametric aims of this book. We are convinced that these new formulations will also be helpful for anybody interested in developing further asymptotic advances on functional nonparametric statistics. The guidelines of this book consist in presenting recent advances for functional variables. However, obtaining asymptotic results needs the use of basic probabilistic tools for real random variables, and many results presented below will concern real random variables.

Section A.1 deals with the notion of almost complete convergence and focuses on the links between this kind of convergence and other more standard ones (such as almost sure convergence or convergence in probability). The statement of almost complete convergence properties relies mainly on some exponential inequality for sums of random variables, and Section A.2 recalls some of these inequalities. Because there are too many in the literature, we concentrate our purpose on those inequalities having a form adapted to the kind of theoretical developments made earlier throughout this book. The last part of this appendix is only useful for people interested in Part IV of this book since focuses on mixing sequences of random variables (either real or functional). More precisely, Section A.3 presents some inequalities for sum of mixing real random variables. Again, as in Section A.2, we have chosen from among the wide literature those inequalities having a form adapted to the framework of this book. Even further, some of these inequalities have been reformulated in new ways, to make their application easier.

It is out of the scope of this book to give the proofs of all these probabilistic tools, and we will mainly refer to the existing literature. However, we will give short proofs for those tools whose proof is not easily accessible in the literature (this is particularly the case in Section A.1 about complete convergence, mainly because the literature on this stochastic mode of convergence is not quite as important). All through this appendix, $(X_n)_{n \in \mathbb{N}}$ and $(Y_n)_{n \in \mathbb{N}}$ are sequences of real random variables, while $(u_n)_{n \in \mathbb{N}}$ is a deterministic sequence of positive real numbers. We will denote also by $(\xi_n)_{n\in\mathbb{Z}}$ a double sequence of random variables (not necessarily real variables), and by $(T_n)_{n\in\mathbb{Z}}$ a double sequence of stationary real random variables. We will use the notation $(Z_n)_{n\in\mathbb{N}}$ for a sequence of independent and centered r.r.v., and $(W_n)_{n\in\mathbb{Z}}$ for a double sequence of stationary dependent and centered r.r.v.

A.1 Almost Complete Convergence

As the reader can see through the proofs presented in this book, the almost complete convergence is in some sense easier to state than the almost sure one. Moreover, as we will see below, this mode of convergence implies other standard modes of convergence. Therefore, because of this double advantage and starting with [C84], it became quite usual for many nonparametricians to express their asymptotic results in terms of complete convergence. Curiously, even though the complete convergence notion was introduced quite a long time ago by [HR47], this notion is not very popular in other statistical communities than nonparametricians. In particular, it is much less popular than the almost sure and the probability stochastic modes of convergence.

We decided to recall some basic definitions and properties about this notion. Because the probabilistic literature about almost complete convergence is not very wide, and because several key properties can be stated in a short and easy way, we decided to give briefly all the proofs of the results stated in this section. We hope that this will be helpful for other purposes than those of this book.

Definition A.1. *One says that $(X_n)_{n\in\mathbb{N}}$ converges almost completely to some r.r.v. X, if and only if*

$$\forall \epsilon > 0, \qquad \sum_{n\in\mathbb{N}} P\left(|X_n - X| > \epsilon\right) \; < \; \infty,$$

and the almost complete convergence of $(X_n)_{n\in\mathbb{N}}$ to X is denoted by

$$\lim_{n\to\infty} X_n = X, \; a.co.$$

This notion, which can sometimes be called more simply *complete convergence* is linked with other stochastic modes of convergence. The first part of Proposition A.2 below will study the link between the almost complete convergence and the convergence in probability. This last mode of convergence will be referred to from now on as convergence p and it is defined by the following property:

$$\lim_{n\to\infty} X_n = X, \text{ p } \Leftrightarrow \forall \epsilon > 0, \lim_{n\to\infty} P\left(|X_n - X| > \epsilon\right) = 0.$$

The second part of Proposition A.2 does the same thing but with the almost sure convergence. This last mode of convergence is defined by the following property:

$$\lim_{n\to\infty} X_n = X, \text{ a.s. } \Leftrightarrow P\left(\lim_{n\to\infty} X_n = X\right) = 1,$$

and it will be referred to from now on as *a.s.* convergence. The reader will find in any elementary probability book a more general presentation of the various links existing between these stochastic modes of convergence and other usual ones (for instance, one can look at [BL87]). The proof of the following property can also be found in [BL87].

Proposition A.2. *If* $\lim_{n\to\infty} X_n = X$, *a.co., then we have:*

i) $\lim_{n\to\infty} X_n = X$, *p,*
ii) $\lim_{n\to\infty} X_n = X$, *a.s.*

Proof. Without loss of generality, we show the result for $X = 0$.

i) This point is obvious.
ii) For all $\epsilon > 0$, we have $\sum_{n\in\mathbb{N}} P\left(|X_n| > \epsilon\right) < \infty$. According to Borel-Cantelli's lemma, it holds that

$$P\left(\overline{\lim}_{n\to\infty}\{|X_n| > \epsilon\}\right) = 0,$$

which can be rewritten as $P\left(A(\epsilon)\right) = 1$ where

$$A(\epsilon) = \{\exists n, \forall m > n, \ |X_m| \leq \epsilon\}.$$

Note that $\left(A(\epsilon)\right)_\epsilon$ is a sequence of embedded events, and so the property $\forall \epsilon > 0, P\left(A(\epsilon)\right) = 1$ implies directly the almost sure convergence, namely:

$$P\left(\forall \epsilon, \exists n, \forall m > n, \ |X_m| \leq \epsilon\right) = 1. \quad \square$$

In the classical literature, the rate of almost sure convergence to 0 for a sequence of r.r.v. is defined by the condition

$$X_n - X = O_{a.s.}\left(u_n\right) \Leftrightarrow P\left(X_n - X = O\left(u_n\right)\right) = 1, \qquad (A.1)$$
$$\Leftrightarrow P\left(\exists C < \infty, \exists n, \forall m > n, |X_m - X| \leq C u_m\right) = 1,$$

while the rate of convergence in probability is defined by

$$X_n - X = O_p(u_n) \Leftrightarrow \lim_m \overline{\lim}_{n \to \infty} P(|X_n - X| < m u_n) = 1.$$

As far as we know, there does not exist such a universally accepted equivalent notion for the complete convergence. The aim of the next definition is to make this notion precise.

Definition A.3. *One says that the rate of almost complete convergence of* $(X_n)_{n \in \mathbb{N}}$ *to* X *is of order* u_n *if and only if*

$$\exists \epsilon_0 > 0, \quad \sum_{n \in \mathbb{N}} P(|X_n - X| > \epsilon_0 u_n) < \infty,$$

and we write

$$X_n - X = O_{a.co.}(u_n).$$

The reader can find other points of view concerning the way to quantify such kind of rates of convergence (see for instance [HSV98] and [HV02]). The interest of our new definition is double. First, it has the merit of providing a precise and formal definition which is interesting from a probabilistic point of view since (see Proposition A.4 below) it implies both previous usual notions of O_p and $O_{a.s.}$. Second, it is interesting for statistical purposes since it turns out to be easier to prove than some O_p or $O_{a.s.}$ property (at least in many situations, including all those treated before in this book).

Proposition A.4. *Assume that* $X_n - X = O_{a.co.}(u_n)$. *We have:*

i) $X_n - X = O_p(u_n)$,
ii) $X_n - X = O_{a.s.}(u_n)$.

Proof. Without loss of generality, we show the results for $X = 0$.

i) The $O_{a.co.}$ definition allows us to write that:

$$\exists m_0, \forall m > m_0, \sum_n P(|X_n| > m u_n) < \infty,$$

and Borel-Cantelli's lemma allows us to get

$$\exists m_0, \forall m > m_0, P\left(\overline{\lim}_{n \to \infty}\{|X_n| > m u_n\}\right) = 0.$$

By applying now Fatou's lemma, one gets

$$\exists m_0, \forall m > m_0, \overline{\lim}_{n \to \infty} P(|X_n| > m u_n) = 0.$$

This is the same as

$$\exists m_0, \forall m > m_0, \underline{\lim}_{n \to \infty} P\left(|X_n| \le m u_n\right) = 1,$$

which implies directly that $X_n = O_p(u_n)$.

ii) As before, by applying Borel-Cantelli's lemma it holds that

$$P\left(\overline{\lim}_{n \to \infty}\{|X_n| > \epsilon_0 u_n\}\right) = 0.$$

This can also be written as

$$\exists \epsilon_0, \ P\left(\exists n, \ \forall m > n, \ |X_m| \le \epsilon_0 u_m\right) = 1,$$

and it comes directly that $X_n = O_{a.s.}(u_n)$. $\square$

Now, we give in Proposition A.5 some elementary calculus rules concerning this stochastic mode of convergence. These elementary rules were used implicitly earlier throughout this book. To conclude this presentation, we will present in Proposition A.6 some more specific results that were also used several times before in this book.

Proposition A.5. *Assume that* $\lim_{n \to \infty} u_n = 0$, $\lim_{n \to \infty} X_n = l_X$, *a.co. and* $\lim_{n \to \infty} Y_n = l_Y$, *a.co., where* l_X *and* l_Y *are two deterministic real numbers.*

    i) We have:
        a) $\lim_{n \to \infty} X_n + Y_n = l_X + l_Y$, *a.co.;*
        b) $\lim_{n \to \infty} Y_n X_n = l_Y l_X$, *a.co.;*
        c) $\lim\limits_{n \to \infty} \dfrac{1}{Y_n} = \dfrac{1}{l_Y}$, *a.co. as long as* $l_Y \ne 0$.
    ii) If $X_n - l_X = O_{a.co.}(u_n)$ *and* $Y_n - l_Y = O_{a.co.}(u_n)$, *we have:*
        a) $(X_n + Y_n) - (l_X + l_Y) = O_{a.co.}(u_n)$;
        b) $X_n Y_n - l_X l_Y = O_{a.co.}(u_n)$;
        c) $\dfrac{1}{Y_n} - \dfrac{1}{l_Y} = O_{a.co.}(u_n)$ *as long as* $l_Y \ne 0$.

Proof. i.a). This proof is obvious since we have

$$P\left(|X_n + Y_n - (l_X + l_Y)| > \epsilon\right) \le P\left(|X_n - l_X| > \frac{\epsilon}{2}\right) + P\left(|Y_n - l_Y| > \frac{\epsilon}{2}\right).$$

ii.a). Applying the last inequality with $\epsilon = \epsilon_0 u_n$ allows us to conclude.

i.b). We can, without loss of generality, only consider the case when $l_X = 0$ (otherwise, just write the decomposition $Y_n X_n = Y_n(X_n - l_X) + Y_n l_X$). So, assume that $l_X = 0$. For $\epsilon < 1/2$ we can write

$$P(|Y_n X_n| > \epsilon) \leq P\left(|Y_n - l_Y| |X_n| > \frac{\epsilon}{2}\right) + P\left(|l_Y X_n| > \frac{\epsilon}{2}\right)$$

$$\leq P\left(|Y_n - l_Y| > \sqrt{\frac{\epsilon}{2}}\right) + P\left(|X_n| > \sqrt{\frac{\epsilon}{2}}\right) + P\left(|l_Y X_n| > \frac{\epsilon}{2}\right)$$

$$\leq P\left(|Y_n - l_Y| > \frac{\epsilon}{2}\right) + P\left(|X_n| > \frac{\epsilon}{2}\right) + P\left(|l_Y X_n| > \frac{\epsilon}{2}\right).$$

The almost complete convergence of $Y_n X_n$ to 0 follows by using the almost complete convergence properties of X_n and Y_n.

ii.b). Applying the last inequality with $\epsilon = \epsilon_0 u_n$ allows us to conclude directly by using the rates of convergence of $X_n - l_X$ and $Y_n - l_Y$.

i.c). The almost complete convergence of Y_n to $l_Y \neq 0$ implies that there exists some $\delta > 0$ (choose for instance $\delta = l_Y/2$) such that

$$\sum_{n \in \mathbb{N}} P\left(|Y_n| \leq \delta\right) < \infty.$$

The proof is performed according to the following steps:

$$P\left(\left|\frac{1}{Y_n} - \frac{1}{l_Y}\right| > \epsilon\right) = P\left(|Y_n - l_Y| > \epsilon |l_Y Y_n|\right)$$

$$\leq P\left(|Y_n - l_Y| > \epsilon |l_Y Y_n| \text{ and } |Y_n| > \delta\right) + P\left(|Y_n| \leq \delta\right)$$

$$\leq P\left(|Y_n - l_Y| > \epsilon \delta |l_Y|\right) + P\left(|Y_n| \leq \delta\right),$$

and the result follows from the almost complete convergence of Y_n to l_Y.

ii.c). This proof follows directly from the last inequality. $\square$

Proposition A.6. *Assume that* $\lim_{n \to \infty} u_n = 0$, $X_n = O_{a.co.}(u_n)$ *and* $\lim_{n \to \infty} Y_n = l_Y$, *a.co., where* l_Y *is a deterministic real number.*
  i) We have $X_n Y_n = O_{a.co.}(u_n)$;
  ii) We have $\dfrac{X_n}{Y_n} = O_{a.co.}(u_n)$ *as long as* $l_Y \neq 0$.

Proof. i.). The almost complete convergence of Y_n to l_Y implies that there exists some $\delta > 0$ such that

$$\sum_{n \in \mathbb{N}} P\left(|Y_n| > \delta\right) < \infty.$$

Now, the proof is performed as follows:

$$\begin{aligned}
P(|Y_n X_n| > \epsilon u_n) &= P(|Y_n X_n| > \epsilon u_n \text{ and } |Y_n| \leq \delta) \\
&+ P(|Y_n X_n| > \epsilon u_n \text{ and } |Y_n| > \delta) \\
&\leq P(|X_n| > \epsilon \delta^{-1} u_n) + P(|Y_n| > \delta).
\end{aligned}$$

So, both of the previous inequalities together with the hypothesis that $X_n = O_{a.co.}(u_n)$ are enough to show that $X_n Y_n = O_{a.co.}(u_n)$.

Proof of ii.). It is a direct consequence of part i) of this proposition together with part i.c) of Proposition A.5. $\square$

A.2 Exponential Inequalities for Independent r.r.v.

In all this section $Z_1, \ldots, Z_n$ will be independent r.r.v. with zero mean. As can be seen throughout this book, the statement of almost complete convergence properties needs to find an upper bound for some probabilities involving sum of r.r.v such as

$$P\left(|\sum_{i=1}^{n} Z_i| > \epsilon \right),$$

where, eventually, the positive real ϵ decreases with n. In this context, there exist powerful probabilistic tools, generically called *Exponential Inequalities*. The literature contains various versions of exponential inequalities. These inequalities differ according to the various hypotheses checked by the variables Z_i's. We focus here on the so-called Bernstein's inequality. This choice was made because the form of Bernstein's inequality is the easiest for the theoretical developments on functional statistics that have been stated throughout our book. Other forms of such exponential inequality can be found in [FN71] (see also [N97] and [N98]).

Proposition A.7. *Assume that*

$$\forall m \geq 2, |\mathbb{E} Z_i^m| \leq (m!/2)(a_i)^2 b^{m-2},$$

and let $(A_n)^2 = (a_1)^2 + \cdots + (a_n)^2$. *Then, we have:*

$$\forall \epsilon \geq 0, \ P\left(\left| \sum_{i=1}^{n} Z_i \right| > \epsilon A_n \right) \leq 2 \exp\left\{ -\frac{\epsilon^2}{2\left(1 + \frac{\epsilon b}{A_n}\right)} \right\}.$$

Such a result being standard, the proof is not given here. Previous proofs of Proposition A.7 below are given in [U37] or [B46]. The reader will also find the proof of some more general results in [Y76] (as well as in many other probability books).

Note that this inequality is stated for non-identically distributed r.r.v. Note also that each variable Z_i may depend on n. Indeed, for our statistical purpose, the next Corollary A.8 is used more often than the previous general proposition.

Corollary A.8. *i) If $\forall m \geq 2$, $\exists C_m > 0$, $\mathbb{E}|Z_1^m| \leq C_m \, a^{2(m-1)}$, we have*

$$\forall \epsilon \geq 0, \; P\left(\left|\sum_{i=1}^{n} Z_i\right| > \epsilon n\right) \leq 2 \exp\left\{-\frac{\epsilon^2 \, n}{2 \, a^2 \, (1+\epsilon)}\right\}.$$

ii) Assume that the variables depend on n (that is, assume that $Z_i = Z_{i,n}$). If $\forall m \geq 2$, $\exists C_m > 0$, $\mathbb{E}|Z_1^m| \leq C_m \, a_n^{2(m-1)}$ and if $u_n = n^{-1} a_n^2 \log n$ verifies $\lim_{n\to\infty} u_n = 0$, we have:

$$\frac{1}{n} \sum_{i=1}^{n} Z_i = O_{a.co.}(\sqrt{u_n}).$$

Proof. i) It suffices to apply Proposition A.7 with $A_n = a\sqrt{n}$ and $b = a^2$.
ii) Taking $\epsilon = \epsilon_0 \sqrt{u_n}$ in the result i), and because the sequence u_n tends to zero, we obtain for some $C' > 0$:

$$P\left(\frac{1}{n}\left|\sum_{i=1}^{n} Z_i\right| > \epsilon_0 \sqrt{u_n}\right) \leq 2 \exp\left\{-\frac{\epsilon_0^2 \log n}{2\left(1 + \epsilon_0 \sqrt{u_n}\right)}\right\} \leq 2 n^{-C' \epsilon_0^2}.$$

and ii) follows directly by choosing ϵ_0 large enough. $\square$

Note also that all previous inequalities are given for unbounded random variables, which is useful for functional nonparametric regression (see Section 6.2.1 and Section 6.3.1). Of course, they apply directly for bounded variables, such as those appearing along functional conditional density or c.d.f. studies (see Chapter 6). This is the reason why we decided to conclude this presentation with a new version of Corolary A.8 which is directly adapted to bounded variables.

Corollary A.9. *i) If* $\exists M < \infty, |Z_1| \leq M$, *and denoting* $\sigma^2 = \mathbb{E}Z_1^2$, *we have*

$$\forall \epsilon \geq 0, \; P\left(\left|\sum_{i=1}^{n} Z_i\right| > \epsilon n\right) \leq 2\exp\left\{-\frac{\epsilon^2 n}{2\sigma^2\left(1 + \epsilon\frac{M}{\sigma^2}\right)}\right\}.$$

ii) Assume that the variables depend on n *(that is,* $Z_i = Z_{i,n}$*) and are such that* $\exists M = M_n < \infty, |Z_1| \leq M$ *and define* $\sigma_n^2 = \mathbb{E}Z_1^2$. *If* $u_n = n^{-1}\sigma_n^2 \log n$ *verifies* $\lim_{n \to \infty} u_n = 0$, *and if* $M/\sigma_n^2 < C < \infty$, *then we have:*

$$\frac{1}{n}\sum_{i=1}^{n} Z_i = O_{a.co.}(\sqrt{u_n}).$$

Proof. i) It suffices to apply Proposition A.7 with $a_i^2 = \sigma^2$, $A_n^2 = n\sigma^2$ and $b = M$.

ii) Because the sequence $v_n = \frac{Mu_n}{\sigma_n^2}$ tends to zero, by choosing $\epsilon = \epsilon_0\sqrt{u_n}$ in the result i) we obtain directly:

$$P\left(\frac{1}{n}\left|\sum_{i=1}^{n} Z_i\right| > \epsilon_0\sqrt{u_n}\right) \leq 2\exp\left\{-\frac{\epsilon_0^2 \log n}{2\left(1 + \epsilon_0\sqrt{v_n}\right)}\right\} \leq 2n^{-C'\epsilon_0^2}.$$

and ii) follows directly by choosing ϵ_0 large enough. $\square$

A.3 Inequalities for Mixing r.r.v.

Nonparametric statistics for real-valued mixing processes have received a lot of attention during the last few decades (see Part IV of this book for a wide scope of references), and it turns out that these statistical advances have been linked with developments of probabilistic tools for mixing sequences. Basically, there are two main kinds of tools that are used for nonparametric purposes: covariance inequalities and exponential inequalities. The aim of this section is to recall some inequalities of these two kinds for real valued α-mixing processes. It is worth noticing that, because of Proposition 10.4 above, these inequalities for real variables will be useful for variables valued in semi-metric spaces. This was done systematically all through Part IV of the book.

Let us first start with some covariance inequality. There is a wide literature concerning covariance inequalities for mixing variables. These inequalities differ both from the kind of mixing condition which is introduced and from the kind of assumptions checked by the variables. The reader will find nice overviews on such results in [RI87] or [Y92]. Shorter reviews are also provided by Chapter 1 of [Y94] and Chapter 1 of [R00]. Such inequalities are also called moment inequalities (see for instance [K94], [CK95] or [EQV02] for recent extensions to higher order moment inequalities useful in nonparametric statistics). We will stay here with α-mixing dependence structures and we just give in the next proposition two covariance inequalities: for bounded and unbounded random variables. As far as we know, the first result below was originally given by [I62] while the second one was previously stated in [D68]. Recall that $(T_n)_{n\in\mathbb{Z}}$ is a stationary sequence of real random variables.

Proposition A.10. *Assume that* $(T_n)_{n\in\mathbb{Z}}$ *is α-mixing. Let us, for some* $k \in \mathbb{Z}$, *consider a real variable* $\mathcal{T}$ *(resp.* $\mathcal{T}'$*) which is* $\mathcal{A}_{-\infty}^{k}$*-measurable (resp.* $\mathcal{A}_{n+k}^{+\infty}$*-measurable).*

i) *If* $\mathcal{T}$ *and* $\mathcal{T}'$ *are bounded, then:*
$$\exists C, 0 < C < +\infty,\ cov(\mathcal{T}, \mathcal{T}') \leq C\alpha(n).$$
ii) *If, for some positive numbers* p, q, r *such that* $p^{-1} + q^{-1} + r^{-1} = 1$, *we have* $E\mathcal{T}^p < \infty$ *and* $E\mathcal{T}'^q < \infty$, *then:*
$$\exists C, 0 < C < +\infty,\ cov(\mathcal{T}, \mathcal{T}') \leq C \left(E\mathcal{T}^p\right)^{\frac{1}{p}} \left(E\mathcal{T}'^q\right)^{\frac{1}{q}} \alpha(n)^{\frac{1}{r}}.$$

Let us now present some exponential inequalities for partial sums of a sequence $(W_n)_{n\in\mathbb{Z}}$ of stationary and centered mixing real random variables. In some sense, even if the forms are not completely comparable, the results presented below are dependent extensions of those described in Section A.2. During the twenty past years, the literature on exponential probabilistic inequalities for mixing sequences was directly linked with the advances on nonparametric statistics for dependent data. This connexion started with the previous Bernstein's type inequalities provided by [B75] and [C84] in a quite more restrictive mixing dependence structure than the α-mixing one. As far as we know, the first exponential inequality for α-mixing variables is due to [C83]. This previous result has been improved in several further works and the reader will find in [B93] and [Ro96] a wide discussion on the bibliography at this point. For our nonparametric functional purpose, we decided to use a dependent version of the Fuk-Nagaev's inequality which was previously introduced by [FN71] for independent variables and refined in [N97] and [N98]. This inequality is recalled in Proposition A.11. For the reasons discussed in Section A.2, we give a result for bounded and one for unbounded variables. To save space, we will state the result without specifying the exact expressions of the constant terms involved in the bounds. Let us introduce the notation:

$$s_n^2 = \sum_{i=1}^{n} \sum_{j=1}^{n} |\text{Cov}(W_i, W_j)|.$$

Proposition A.11. *Assume that* $(W_n)_{n \in \mathbb{N}^*}$ *are identically distributed and are arithmetically α-mixing with rate $a > 1$.*

i) If there exist $p > 2$ and $M > 0$ such that $\forall t > M, P(|W_1| > t) \leq t^{-p}$, then we have for any $r \geq 1$ and $\epsilon > 0$ and for some $C < \infty$:

$$P\left(\left|\sum_{i=1}^{n} W_i\right| > \epsilon\right) \leq C\left\{\left(1 + \frac{\epsilon^2}{r\,s_n^2}\right)^{-r/2} + n\,r^{-1}\left(\frac{r}{\epsilon}\right)^{(a+1)p/(a+p)}\right\}.$$

ii) If there exist $M < \infty$ such that $|W_1| \leq M$, then we have for any $r \geq 1$ and for some $C < \infty$:

$$P\left(\left|\sum_{i=1}^{n} W_i\right| > \epsilon\right) \leq C\left\{\left(1 + \frac{\epsilon^2}{r\,s_n^2}\right)^{-r/2} + n\,r^{-1}\left(\frac{r}{\epsilon}\right)^{a+1}\right\}.$$

The result of Proposition A.11 is stated and proved in [R00] in a more general framework than ours. To make the use of such a probabilistic result easier, both for us in this book and for anybody else who could be interested in developing further advances, we propose new formulations of this inequality (see Corollary A.12 and Corollary A.13) which are directly adapted to nonparametric statistics applications. Both corollaries differ with the kind of mixing coefficients: arithmetic or geometric.

Corollary A.12. *Assume that the variables depend on n (that is, $W_i = W_{i,n}$), and that $W_1, \ldots W_n$ are n successive terms of a mixing sequence with arithmetic coefficients of order $a > 1$. Let us consider the deterministic sequence $u_n = n^{-2}s_n^2 \log n$. Assume also that one among both sets of assumptions is satisfied:*

i) $\exists p > 2, \exists \theta > 2, \exists M = M_n < \infty$, such that
$$\forall t > M_n, P(|W_1| > t) \leq t^{-p} \text{ and } s_n^{-\frac{(a+1)p}{a+p}} = o(n^{-\theta}),$$

or

ii) $\exists M = M_n < \infty, \exists \theta > 2$ such that
$$|W_1| \leq M_n \text{ and } s_n^{-(a+1)} = o(n^{-\theta}).$$

Then we have
$$\frac{1}{n}\sum_{i=1}^{n} W_i = O_{a.co.}(\sqrt{u_n}).$$

Proof. Note first that the conditions on s_n insure that $\lim_{n \to \infty} u_n = 0$. We will prove both assertions at once by using the notation:

$$q = \begin{cases} (a+1)p/(a+p) \text{ under the conditions of i)} \\ a+1 \text{ under the conditions of ii).} \end{cases}$$

Take $r = (\log n)^2$ and apply Proposition A.11, to get for any $\epsilon_0 > 0$:

$$P\left(\left|\frac{1}{n}\sum_{i=1}^{n} W_i\right| > \epsilon_0\sqrt{u_n}\right)$$

$$\leq C\left(1 + \frac{\epsilon_0^2 n^2 u_n}{(\log n)^2 s_n^2}\right)^{-\frac{(\log n)^2}{2}} + n(\log n)^{-2}\left(\frac{\log n}{n\epsilon_0\sqrt{u_n}}\right)^q$$

$$\leq C\left(1 + \frac{\epsilon_0^2}{\log n}\right)^{-\frac{(\log n)^2}{2}} + n(\log n)^{-2}\left(\frac{\sqrt{\log n}}{\epsilon_0 s_n}\right)^q.$$

Using the fact that $\log(1+x) = x - x^2/2 + o(x^2)$ when $x \to 0$, we get:

$$P\left(\left|\frac{1}{n}\sum_{i=1}^{n} W_i\right| > \epsilon_0\sqrt{u_n}\right) \leq Ce^{-\frac{\epsilon_0^2 \log n}{2}} + n(\log n)^{-2+\frac{q}{2}} s_n^{-q}\epsilon_0^{-q}.$$

Finally, the condition on s_n allows us to get directly that there exist some ϵ_0 and some $\eta_0 > 0$ such that

$$P\left(\left|\frac{1}{n}\sum_{i=1}^{n} W_i\right| > \epsilon_0\sqrt{u_n}\right) \leq Cn^{-1-\eta_0}, \qquad (A.-13)$$

and this proof is complete. □

Corollary A.13. *Assume that the variables depend on n (that is, $W_i = W_{i,n}$), and that $W_1, \ldots W_n$ are n successive terms of a mixing sequence with geometrical coefficients. Let us consider the deterministic sequence $u_n = n^{-2}s_n^2 \log n$. Assume that one of these two sets of assumptions is satisfied:*

i) $\exists p > 2, \exists \theta > \frac{2}{p}, \exists M = M_n < \infty$, such that
$\forall t > M_n, P(|W_1| > t) \leq t^{-p}$ and $s_n^{-1} = o(n^{-\theta})$,

or

ii) $\exists M = M_n < \infty, \exists \theta > 2$ such that
$|W_1| \leq M_n$ and $s_n^{-1} = o(n^{-\theta})$.

Then we have

$$\frac{1}{n}\sum_{i=1}^{n} W_i = O_{a.co.}(\sqrt{u_n}).$$

Proof. This proof is obvious since the geometrical decaying of the coefficients allows us to apply Corollary A.12 for any value of a. □

References

[ABH03] Abdous, B., Berlinet, A., Hengartner, N. A general theory for kernel estimation of smooth functionals of the distribution function and their derivatives. *Rev. Roumaine Math. Pures Appl.*, **48**, 217-232 (2003).

[ABC03] Abraham, C., Biau, G., Cadre, B. Simple estimation of the mode of a multivariate density. *Canad. J. Statist.*, **31**, 23-34 (2003).

[ABC04] Abraham, C., Biau, G., Cadre, B. On the asymptotic properties of a simple estimation of the mode. *ESAIM Proba. Statist.*, **8**, 1-11 (2004).

[ABC05] Abraham, C., Biau, G., Cadre, B. On the kernel rule for function classification. *Ann. Inst. Statist. Math.*, to appear (2005).

[ACMM03] Abraham, C., Cornillon, P., Matzner-Löber, E., Molinari, N. Unsupervised curve clustering using B-splines. *Scand. J. Statist.*, **30**, 581-595 (2003).

[AOV99] Aguilera, A., Ocaña, F., Valderrama, M. Stochastic modelling for evolution of stock prices by means of functional PCA. *Applied Stochastic Models in Business and Industry*, **15**, 227-234 (1999).

[AOV99b] Aguilera, A., Ocaña, F., Valderrama, M. Forecasting with unequally spaced data by a functional principal component approach. *Test*, **8**, 233-253 (1999).

[AFK05] Ait Saidi, A., Ferraty, F., Kassa, R. Single functional index model for time series. *Rev. Roumaine Math. Pures Appl.*, **50** (4), 321-330 (2005).

[AP03] Akritas, M., Politis, D. (ed.) *Recent advances and trends in nonparametric statistics*. Elsevier, Amsterdam (2003).

[ACEV04] Aneiros-Pérez, G., Cardot, H., Estevez-Pérez, G., Vieu, P. Maximum ozone forecasting by functional nonparametric approaches. *Environmetrics*, **15**, 675-685 (2004).

[AV04] Aneiros-Pérez, G., Vieu, P. Semi-functional partial linear regression. *Statist. Prob. Lett.*, in print (2005).

[AR99] Arteche, J., Robinson, P. Seasonal and cyclical long memory. *Asymptotics, nonparametrics, and time series*, 115-148, Statist. Textbooks Monogr., **158**, Dekker, New York (1999).

[AT90] Auestad, B., Tjostheim, D. Identification of nonlinear time series: first order characterization and order determination. *Biometrika*, **77** (4), 669-687 (1990).

[A81] Azzalini, A. A note on the estimation of a distribution function and quantiles by kernel method. *Biometrika*, **68**, 326-328 (1981).

240   References

[B78]   Banon, G. Nonparametric identification for diffusion processes. *Siam. Journ. Control and Optim.*, **16**, 380-395 (1978).

[BH01] Bashtannyk, D., Hyndman, R. Bandwidth selection for kernel conditional density estimation. *Comput. Statist. Data Anal.*, **36**, 279-298 (2001).

[BCW88] Becker, R., Chambers, J., Wilks, A. *The New S Language.* Wadsworth & Brooks/Cole, Pacific Grove (1988)

[BL02] Belinsky, E., Linde, W. Small ball probabilities of fractional Brownian sheets via fractional integration operators. *J. Theoret. Probab.*, **15** (3), 589-612 (2002).

[B93]   Berlinet, A. Hierarchies of higher order kernels. *Proba. Theory Related Fields* **94**, 489-504 (1993).

[BGM01] Berlinet, A., Gannoun, A., Matzner-Lober, E. Asymptotic normality of convergent estimates of conditional quantiles. *Statistics*, **35**, 139-169 (2001).

[BL01] Berlinet, A., Levallois, S. Higher order analysis at Lebesgue points. In M. Puri (ed.) *Asymptotics in Statistics and Probability*, VSP (2001).

[BT04] Berlinet, A., Thomas-Agnan, C. *Reproducing Kernel Hilbert Spaces in Probability and Statistics*, Kluwer Academic Publishers, Dordrecht (2004).

[B46]   Bernstein, S. *Probability Theory, 4th ed.* (in russian), ed. M.L. Gostechizdat (1946).

[B03]   Bigot, J. Automatic landmark registration of 1D curves. In: M. Akritas and D. Politis (eds.), *Recent advances and trends in nonparametric statistics*. Elsevier, Amsterdam, 479-496 (2003).

[BF95] Boente, G., Fraiman, R. Asymptotic distribution of data-driven smoothers in density and regression estimation under dependence. *Canad. J. Statist.*, **23** (4), 383-397 (1995).

[BF95b] Boente, G., Fraiman, R. Asymptotic distribution of smoothers based on local means and local medians under dependence. *Multivariate Anal.*, **54** (1), 77-90 (1995).

[BF00] Boente, G., Fraiman, R. Kernel-based functional principal components. *Statistics and Probability Letters*, **48**, 335-345 (2000)

[B99]   Bogachev, V.I. *Gaussian measures.* Math surveys and monographs, **62**, Amer. Math. Soc. (1999).

[BT92] Borggaard, C., Thodberg, H.H. Optimal Minimal Neural Interpretation of Spectra. *Analytical Chemistry* **64**, 545-551 (1992).

[B75]   Bosq, D. Inégalité de Bernstein pour un processus mélangeant (in french). *Compte Rendus Acad. Sci. Paris*, Ser. A, **275**, 1095-1098 (1975).

[B91]   Bosq, D. Modelization, nonparametric estimation and prediction for continuous time processes. In: G. Roussas (Ed) *Nonparametric functional estimation and related topics*. NATO ASI Series **335**, Kluwer Acad. Publ., Dordrecht, 509-530 (1991).

[B93]   Bosq, D. Bernstein's type large deviation inequality for partial sums of strong mixing process. *Statistics*, **24**, 59-70 (1993).

[B98]   Bosq, D. *Nonparametric Statistics for Stochastic Process. Estimation and Prediction.* 2nd edition, Lecture Notes in Statistics, **110**, Springer-Verlag, New York (1998).

[B00]   Bosq, D. *Linear Processes in Function Spaces, Theory and Applications.* Lecture Notes in Statistics, 149, Springer-Verlag, New York (2000).

[B03]   Bosq, D. Berry-Esseen inequality for linear processes in Hilbert spaces. *Statist. Probab. Lett.*, **63** (3), 243-247 (2003).

[B05]   Bosq, D. *Inférence et prévision en grandes dimensions* (in french). Economica (2005).

[BD85]   Bosq, D., Delecroix, M. Nonparametric prediction of a Hilbert-space valued random variable. *Stochastic Process. Appl.*, **19** (2), 271-280 (1985).

[BL87]   Bosq, D., Lecoutre, J.P. *Théorie de l'estimation fo.nctionnelle* (in french). Economica (1987).

[B86]   Bradley, R. Basic properties of strong mixing conditions. In *Dependence in probability and statistics: a survey of recent results*. Birkhauser, Boston (1986).

[C01]   Cadre, B. Convergent estimators for the L_1 median of a Banach valued random variable. *Statistics*, **35**, 509-521 (2001).

[Ca02]   Cadre, B. Contributions à l'étude de certains phénomènes stochastiques (in french). Habilitation à Diriger des Recherches, Université Montpellier II, (2002).

[C91]   Cai, Z. Strong consistency and rates for recursive nonparametric conditional probability density estimates under (α, β)-mixing conditions. *Stochastic Process. Appl.*, **38** (2), 323-333 (1991).

[C02]   Cai, Z. Regression quantiles for time series. *Econometric Theory*, **18** (1), 169-192 (2002).

[CSV00] Camlong-Viot, Ch., Sarda, P., Vieu, Ph. Additive time series: the kernel integration method. *Math. Methods Statist.*, **9** (4), 358-375 (2000).

[C83]   Carbon, M. Inégalité de Bernstein pour les processus fortement mélangeants non nécessairement stationnaires (in french). *Compte Rendus Acad. Sci. Paris*, Ser. A, **297**, 303-306 (1983).

[CD93]   Carbon M., Delecroix, M. Nonparametric forecasting in time series: a computational point of view. *Applied stochastic models and data analysis*, **9**, 215-229 (1993).

[C04]   Cardot, H. Contributions à la modélisation statistique fonctionnelle (in french). Habilitation à Diriger des Recherches, Université Paul Sabatier, Toulouse (2004).

[CCS04] Cardot, H., Crambes, C., Sarda, P. Estimation spline de quantiles conditionnels pour variables explicatives fonctionnelles (in french). *Compte Rendus Acad. Sci. Paris*, **339**, 141-144 (2004).

[CFG03] Cardot, H., Faivre, R., Goulard, M. Functional approaches for predicting land use with the temporal evolution of coarse resolution remote sensing data. *J. Applied Statist.*, **30**, 1185-1199 (2003).

[CFMS03] Cardot, H., Ferraty, F., Mas, A., Sarda, P. Testing hypotheses in the functional linear model. *Scand. J. of Statist.*, **30**, 241-255 (2003).

[CFS99] Cardot, H., Ferraty, F., Sarda, P. Linear Functional Model. *Statist. & Prob. Letters*, **45**, 11-22 (1999).

[CFS03] Cardot, H., Ferraty, F., Sarda, P. Spline Estimators for the Functional Linear Model. *Statistica Sinica.*, **13**, 571-591 (2003).

[CGS04] Cardot, H., Goia, A., Sarda, P. Testing for no effect in functional linear regression models, some computational approaches. *Comm. Statist. Simulation Comput.* **33**, 179-199 (2004).

[CLS86] Castro, P.E., Lawton, W.H., Sylvestre, E.A. Principal modes of variation for processes with continuous sample curves. *Technometrics*, **28**, 329-337.

[C98]   Chambers, J. *Programming with Data: A Guide to the S Language*. Springer-Verlag, New York (1998)

242   References

[C97]   Chen, G. Berry-Essen-type bounds for the kernel estimator of conditional distribution and conditional quantiles. *J. Statist. Plann. Inference*, **60**, 311-330 (1997).

[CL03]   Chen, X., Li, W. Quadratic functionals and small ball probabilities for the m-fold integrated Brownian motion. *Ann. Probab.*, **31** (2), 1052-1077 (2003).

[C95]   Chu, C. Bandwidth selection in nonparametric regression with general errors. *J. Statist. Plann. Inference*, **44** (3), 265-275 (1995).

[C94]   Chu, C. Binned modified cross-validation with dependent errors. *Comm. Statist. Theory Methods*, **23** (12), 3515-3537 (1994).

[CFGR05] Clarkson, D.B., Fraley, C., Gu, C.C., Ramsay, J.S. *S+ functional data analysis user's guide*. Comp. Statist. Series, Springer, New York (2005).

[C76]   Collomb, G. Estimation nonparamétrique de la régression (in french). PhD Université Paul Sabatier, Toulouse (1976).

[C80]   Collomb, G. Estimation nonparamétrique de probabilités conditionnelles (in french). *Compte Rendus Acad. Sci. Paris*, Ser. A, **291**, 303-306 (1980).

[C84]   Collomb, G. Propriétés de convergence presque complète du prédicteur à noyau (in french). *Z. Wahrscheinlichkeitstheorie verw. Gebiete*, **66**, 441-460 (1984).

[C85]   Collomb, G. Nonparametric regression: an up-to-date bibliography. *Statistics*, **2**, 309-324 (1985).

[CH86]   Collomb, G., Härdle, W. Strong uniform convergence rates in robust nonparametric time series analysis and prediction: kernel regression estimation from dependent observations. *Stochastic Process. Appl.*, **23** (1), 77-89 (1986).

[CHH87]   Collomb, G., Härdle, W., Hassani, S. A note on prediction via conditional mode estimation. *J. Statist. Plann. and Inf.*, **15**, 227-236 (1987).

[CK95]   Cox, D., Kim, T.Y. Moment bounds for mixing random variables useful in nonparametric function estimation. *Stochastic Process Appl.*, **56**, (1), 151-158 (1995).

[Co04]   Comte, F. Kernel deconvolution of stochastic volatility models. *J. Time Ser. Anal.*, **25** (4), 563-582 (2004).

[CF06]   Cuesta-Albertos, J. Fraiman, R. Impartial Trimmed k-Means and Classification Rules for Functional Data. Preprint (2006).

[CFF02]   Cuevas, A., Febrero, M., Fraiman, R. Linear functional regression: the case of fixed design and functional response. *Canad. J. of Statist.*, **30** (2), 285-300 (2002).

[CFF04]   Cuevas, A. Febrero, M., Fraiman, R. An anova test for functional data. *Comput. Statist. Data Anal.*, **47**, 111-122 (2004).

[D02]   Dabo-Niang, S. Sur l'estimation de la densité en dimension infinie: applications aux diffusions (in french). PhD Paris VI (2002).

[D04]   Dabo-Niang, S. Density estimation by orthogonal series in an infinite dimensional space: application to processes of diffusion type I. *J. Nonparametr. Stat.*, **16**, 171-186 (2004).

[D04b]   Dabo-Niang, S. Kernel density estimator in an infinite-dimensional space with a rate of convergence in the case of diffusion process. *Appl. Math. Lett.*, **17**, 381-386 (2004).

[DFV04]   Dabo-Niang, S., Ferraty, F., Vieu P. Nonparametric unsupervised classification of satellite wave altimeter forms. In *Proceedings in Computational Statistics*, ed. J. Antoch, Physica-Verlag, Heidelberg, 879-886 (2004).

[DFV06] Dabo-Niang, S., Ferraty, F., Vieu, P. Mode estimation for functional random variable and its application for curves classification. *Far East J. Theor. Stat.*, **18**, 93-119 (2006).

[D02] Dabo-Niang, S., Rhomari, N. Estimation non paramétrique de la régression avec variable explicative dans un espace métrique (in french). *Compte Rendus Acad. Sci. Paris*, **336**, 75-80 (2003).

[DG02] Damon, J., Guillas, S. The inclusion of exogenous variables in functional autoregressive ozone forecasting. *Environmetrics*, **13**, 759-774 (2002).

[DG05] Damon, J., Guillas, S. Estimation and simulation of autoregressive Hilbertian processes with exogenous variables. *Statistical Inference for Stochastic Processes*, **8**, 185-204 (2005).

[DPR82] Dauxois, J., Pousse, A., Romain, Y. Asymptotic theory for the principal component analysis of a random vector function: some application to statistical inference. *J. Multivariate Anal.*, **12**, 136-154 (1982).

[D68] Davydov, Y. Convergence of distributions generated by stationary stochastic processes. *Theory of Probability and its Applications*, **13**, 691-696 (1968).

[deB78] de Boor, C. A practical guide to splines. Springer, New York (1978).

[deGZ03] de Gooijer, J., Zerom, D. On conditional density estimation. *Statist. Neerlandica*, **57**, 159-176 (2003).

[D03] Dereich, S. Small ball probabilities around random centers of Gaussian measures and applications to quantization. *J. Theoret. Probab.*, **16** (2), 427-449 (2003).

[DFMS03] Dereich, S., Fehringer, F., Matoussi, A., Scheutzow, M. On the link between small ball probabilities and the quantization problem for Gaussian measures on Banach spaces. *J. Theoret. Probab.*, **16** (1), 249-265 (2003).

[DGL96] Devroye, L., Györfi, L., Lugosi, G. *A probabilistic theory of pattern recognition.* Springer-Verlag, New York (1996).

[D95] Doukhan, P. *Mixing: Properties and examples.* Lecture Notes in Statistics, Springer-Verlag, New York, **85**, (1995).

[DM01] Ducharme, G., Mint El Mouvid, M. Convergence presque sûre de l'estimateur linéaire local de la fonction de répartition conditionnelle (in french). *C. R. Acad. Sci. Paris*, **333** 873-876 (2001).

[E80] Eddy, W. Optimum kernel estimators of the mode. *Ann. Statist.*, **8**, 870-882 (1980).

[E82] Eddy, W. The asymptotic distributions of kernel estimators of the mode. *Z. Wahrsch. Verw. Gebiete*, **59**, 279-290 (1982).

[EAV05] Escabias, M., Aguilera, A., Valderrama, M. Modeling environmental data by functional principal component logistic regression. *Environmetrics*, **16**, 95-107 (2005).

[EQV02] Estèvez-Pèrez, G., Quintela-del-Rio, A., Vieu, P. Convergence rate for cross-validatory bandwidth in kernel hazard estimation from dependent samples. *J. Statist. Plann. Inference*, **104** (1), 1-30 (2002).

[EV03] Estévez, G., Vieu, P. Nonparametric estimation under long memory dependence. *J. Nonparametr. Stat.*, **15** (4-5), 535-551 (2003).

[EO05] Ezzahrioui, M., Ould-Saïd, E. Asymptotic normality of nonparametric estimators of the conditional mode function for functional data. Preprint (2005).

[EO05b] Ezzahrioui, M., Ould-Saïd, E. Asymptotic normality of the kernel estimators of the conditional quantile in the normed space. Preprint (2005).

[F85]   Faden, A. The existence of regular conditional probabilities: necessary and sufficient conditions. *Ann. Probab.* **13**, 288-298 (1985).

[FG96]   Fan, J., Gijbels, I. Local polynomial fitting. In: M. Schimek (Ed) *Smoothing and regression; Approaches, Computation, and Application.* Wiley Series in Probability and Statistics, Wiley, New York, 229-276 (2000).

[FG00]   Fan, J., Gijbels, I. *Local polynomial modelling and its applications.* Chapman and Hall, London (1996).

[FL98]   Fan, J., Lin, S. Test of Significance when data are curves. *Jour. Ameri. Statist. Assoc.*, **93**, 1007-1021 (2004).

[FY03]   Fan, J., Yao, Q. *Non linear time series. Nonparametric and parametric methods.* Springer series in Statistics. Springer-Verlag, New York (2003).

[FYT96] Fan, J., Yao, Q., Tong, H. Estimation of conditional densities and sensitivity measures in nonlinear dynamical systems. *Biometrika*, **83**, 189-206 (1996).

[FY04]   Fan, J., Yim, T.H. A data-driven method for estimating conditional densities. *Biometrika*, **91**, 819-834 (2004).

[FZ00]   Fan, J., Zhang, J.T. Two-step estimation of functional linear models with applications to longitudinal data. *J. R. Stat. Soc. Ser. B Stat. Methodol.*, **62**, 303-322 (2000).

[FGG05] Fernández de Castro, B., Guillas, S., Gonzalez Manteiga, W. Functional samples and bootstrap for the prediction of $SO2$ levels. *Technometrics*, **4** (2), 212-222 (2005).

[F03]   Ferraty, F. Modélisation statistique pour variables fonctionnelles: Théorie et Applications (in french). Habilitation à Diriger des Recherches, Université Paul Sabatier, Toulouse (2003).

[FGV02] Ferraty, F., Goia, A., Vieu, P. Functional nonparametric model for time series: a fractal approach to dimension reduction. *TEST*, **11** (2), 317-344 (2002).

[FLV05] Ferraty, F., Laksaci, A., Vieu, P. Estimation of some characteristics of the conditional distribution in nonparametric functional models. *Statistical Inference for Stochastic Processes*, **9** in print (2006).

[FLV05b] Ferraty, F., Laksaci, A., Vieu, P. Functional time series prediction via conditional mode estimation. *Compte Rendus Acad. Sci. Paris*, **340**, 389-392 (2005).

[FMV04] Ferraty, F., Mas, A., Vieu, P. Advances on nonparametric regression from functional data. Preprint (2005).

[FPV03] Ferraty, F., Peuch, A., Vieu, P. Modèle à indice fonctionnel simple (in french). *Compte Rendus Acad. Sci. Paris*, **336**, 1025-1028 (2003).

[FRV05] Ferraty, F., Rabhi, A., Vieu, P. Conditional quantiles for functional dependent data with application to the climatic El niño phenomenon. *Sankhya*, **67**, 378-398 (2005).

[FV00]   Ferraty, F., Vieu, P. Dimension fractale et estimation de la régression dans des espaces vectoriels semi-normés (in french). *Compte Rendus Acad. Sci. Paris*, **330**, 403-406 (2000).

[FV02]   Ferraty, F., Vieu, P. The functional nonparametric model and application to spectrometric data. *Computational Statistics*, **17**, 545-564 (2002).

[FV03]   Ferraty, F., Vieu, P. Curves discrimination: a nonparametric functional approach. *Computational Statistics & Data Analysis*, **44**, 161-173 (2003).

[FV03b] Ferraty, F., Vieu, P. Functional nonparametric statistics: a double infinite dimensional framework. In: M. Akritas and D. Politis (eds.) *Recent advances and trends in nonparametric statistics.* Elsevier, Amsterdam, 61-78 (2003).

[FV04] Ferraty, F., Vieu, P. Nonparametric models for functional data, with applications in regression, time series prediction and curves discrimination. *J. of Nonparametric Statistics,* **16**, 111-127 (2004).

[FM01] Fraiman, R., Muniz, G. Trimmed means for functional data. *Test,* **10**, 419-440 (2001).

[FF93] Frank, I.E., Friedman, J.H. A statistical view of some chemometrics regression tools (with discussions). *Technometrics* **35** (2), 109-148 (1993).

[FS81] Friedman, J., Stuetzle, W. Projection pursuit regression. *J. Amer. Statist. Assoc.* , **76**, 817-823 (1981).

[FN71] Fuk, D.K, Nagaev, S.V. Probability inequalities for sums of independant random variables. *Theory Probab. Appl.,* **16**, 643-660 (1971).

[G90] Gannoun, A. Estimation non paramétrique de la médiane conditionnelle. Application à la prévision. (in french) *C. R. Acad. Sci. Paris Sér. I,* **310** (5), 295-298 (1990).

[G02] Gannoun, A. Sur quelques problèmes d'estimation fonctionnelle: Théorie, méthodes et applications (in french). Habilitation à Diriger des Recherches, Université de Montpellier 2, (2002).

[GSY03] Gannoun, A., Saracco, J., Yu, K. Nonparametric prediction by conditional median and quantiles. *J. Statist. Plann. Inference,* **117** (2), 207-223 (2003).

[GHLT04] Gao, F., Hannig, J., Lee, T.-Y., Torcaso, F. Exact L_2 small balls of Gaussian processes. *J. Theoret. Probab.,* **17** (2), 503-520 (2004).

[GHT03] Gao, F., Hannig, J., Torcaso, F. Integrated Brownian motions and exact $L2$ small balls. *Ann. Probab.,* **31** (3), 1320-1337 (2003).

[G98] Gao, J. Semiparametric regression smoothing of non-linear time series. *.Scand. J. Statist.,* **25** (3), 521-539 (1998).

[GT04] Gao, J., Tong, H. Semiparametric non-linear time series model selection. *J. R. Stat. Soc. Ser. B Stat. Methodol.,* **66** (2), 321-336 (2004).

[GHP98] Gasser, T., Hall, P., Presnell, B. Nonparametric estimation of the mode of a distribution of random curves. *J. R. Statist. Soc. B Stat. Methodol.,* **60**, 681-691 (1998).

[GK95] Gasser, T., Kneip, A. Searching for structure in curve samples. *J. Amer. Statist. Assoc.* , **31**, 1179-1188 (1995).

[GM84] Gasser, T., Müller, H.G. Kernel estimation of regression function. In: T. Gasser and M. Rosenblatt (eds.) *Smoothing techniques for curve estimation.* Springer, Heidelberg, 23-68 (1979).

[GK86] Geladi, P., Kowlaski, B. Partial least square regression: A tutorial, *Analytica Chemica Acta,* **35**, 1-17 (1986).

[G99] Gelfand, A. Approaches for semiparametric Bayesian regression. *Asymptotics, nonparametrics, and time series.* 615-638, Statist. Textbooks Monogr., **158**, Dekker, New York (1999).

[G74] Geffroy, J. Sur l'estimation d'une densité dans un espace métrique. (in french). *C. R. Acad. Sci. Paris Sér. A,* **278**, 1449-1452 (1974).

[Gh99] Ghosh, S. *Asymptotics, nonparametrics, and time series. A tribute to Madan Lal Puri.* Statistics Textbooks and Monographs, **158**. Marcel Dekker, Inc., New York, (1999).

[GC04] Ghosh, A.K., Chaudhuri, P. Optimal smoothing in kernel discriminant analysis. *Statist. Sinica,* **14**, 457-483 (2004).

[GQV802] Gonzalez-Manteiga, W., Quintela-del-Ro, A., Vieu, P. A note on variable selection in nonparametric regression with dependent data. *Statist. Probab. Lett.*, **57** (3), 259-268 (2002).

[GZ03] De Gooijer, J., Zerom, D. On conditional density estimation. *Statist. Neerlandica*, **57**, 159-176. (2003).

[Go99] Gordon, A. D. *Classification.* Chapman & Hall (1999).

[GLP04] Graf, S., Luschgy, H., Pagès, G. Functional quantization and small ball probabilities for Gaussian processes. *J. Theoret. Probab.*, **16** (4), 1047-1062 (2004).

[G02] Guillas, S. Doubly stochastic Hilbertian processes. *Journal of Applied Probability*, **39**, 566-580 (2002).

[G81] Györfi, L. Recent results on nonparametric regression estimate and multiple classification. *Problems Control Inform. Theory*, **10**, 43-52 (1981).

[GHSV89] Györfi, L., Härdle, W., Sarda, P., Vieu, P. *Nonparametric curve estimation from time series.* Lecture Notes in Statistics, **60**. Springer-Verlag, Berlin (1989).

[HH090] Hall, P., Hart, J.D. Nonparametric regression with long-range dependence. *Stochastic Process. Appl.*, **36** (2), 339-351 (1990).

[HH02] Hall, P., Heckman, N. Estimating and depicting the structure of the distribution of random functions, *Biometrika*, **89**, 145-158 (2002).

[HLP95] Hall, P., Lahiri, S., Polzehl, J. On bandwidth choice in nonparametric regression with both short and long range dependent errors. *Ann. Statist.*, **23** (6), 1921-1936 (1995).

[HPP01] Hall, P., Poskitt, P., Presnell, D. A functional data-analytic approach to signal discrimination. *Technometrics*, **35**, 140-143 (2001).

[HWY99] Hall, P., Wolff, R.C., Yao, Q. Methods for estimating a conditional distribution function. *J. Amer. Statist. Assoc.* , **94**, 154-163 (1999).

[H97] Hand, D. J. *Construction and assessment of classification rules.* John Wiley & Sons, Chichester, England (1997).

[H90] Härdle, W. *Applied nonparametric regression.* Cambridge Univ. Press, UK (1990).

[HLG00] Härdle, W., Liang, H., Gao, J. *Partially linear models.* Contributions to Statistics. Physica-Verlag, Heidelberg (2000).

[HM93] Härdle, W., Mammen, E. Comparing nonparametric versus parametric regression fits. *Ann. Statist.* **21**, 1926-1947 (1993).

[HM85] Härdle, W., Marron, J.S. Optimal bandwidth choice in nonparametric regression function estimation. *Ann. of Statist.*, **13**, 1465-1481 (1985).

[HM91] Härdle, W., Marron, J.S. Bootstrap simultaneous errors bars for nonparametric regression. *Ann. of Statist.*, **16**, 1696-1708 (1991).

[HM00] Härdle, W., Müller, M. Multivariate and semiparametric regression. In: M. Schimek (Ed) *Smoothing and regression; Approaches, Computation, and Application.* Wiley Series in Probability and Statistics, Wiley, New York, 357-392 (2000).

[HMSW04] Härdle, W., Müller, M., Sperlich, S., Werwatz, A. *Nonparametric and semiparametric models.* Springer Series in Statistics. Springer-Verlag, New York (2004).

[HV92] Härdle, W., Vieu, P. Kernel regression smoothing of time series. *J. Time Ser. Anal.*, **13** (3), 209-232 (1992).

[H96] Hart, J. Some automated methods of smoothing time-dependent data. *J. Nonparametr. Statist.*, **6** (2-3), 115-142 (1996).

[HV90]   Hart, J., Vieu, P. Data-driven bandwidth choice for density estimation based on dependent data. *Ann. Statist.*, **18** (2), 873-890 (1990).

[H75]   Hartigan, J. *Clustering algorithms.* Wiley, New York (1975).

[HW79]   Hartigan, J., Wong, M. A k-means clustering algorithm. *J. Appl. Statist.*, **28**, 100-108 (1979).

[HBF01] Hastie, T., Buja, A., Friedman, J. *The elements of statistical Learning. Data mining, inference, and prediction.* Springer-Verlag, New York (2001).

[HBT94] Hastie, T., Buja, A., Tibshirani, R. Flexible discriminant analysis by optimal scoring. *J. Amer. Statist. Assoc.* **89**, 1255-1270 (1994)

[HBT95] Hastie, T., Buja, A., Tibshirani, R. Penalized discriminant analysis. *Ann. Statist.* , **13** 435-475 (1995).

[H90]   Helland, I. PLS regression and statistical models, *Scand. J. Statist.* **17**, 97-114 (1990).

[HZ04]   Herrmann, E., Ziegler, K. Rates on consistency for nonparametric estimation of the mode in absence of smoothness assumptions. *Statist. Probab. Lett.*, **68**, 359-368 (2004).

[HK70]   Hoerl, A., Kennard, R. Ridge regression: biased estimation for nonorthogonal problems, *Technometrics* , **8**, 27-51 (1970).

[H75]   Hornbeck, R.W. *Numerical methods.* Quantum Publishers Inc. New York (1975).

[HVZ]   Horova, I., Vieu, P., Zelinka, J. Optimal choice of nonparametric estimates of a density and of its derivatives. *Statistics & Decisions* **20**, 355-378 (2002).

[H88]   Höskuldson, A. PLS regression methods, *J. Chemometrics*, **2**, 211-228 (1988).

[HR47]   Hsu, P., Robbins, H. Complete convergence and the law of large numbers. *Proc. Nat. Acad. Sci. USA*, **33**, 25-31 (1947).

[HSV98] Hu, T.C., Szynal, D., Volodin, A. I. A note on complete convergence for arrays. *Statistics and Probability Letters* , **38**, 27-31 (1998).

[HV02]   Hu, T.C., Volodin, A. Addendum to "A note on complete convergence for arrays". *Statistics and Probability Letters* , **47**, 209-211 (2002).

[H85]   Huber, P. Projection pursuit. *Ann. Statist.* , **13**, 435-475 (1985).

[I62]   Ibragimov., I. Some limit theorems for stationary processes. *Theory of Probability and its Applications* , **7**, 349-362 (1962).

[IM02]   Ioannides, D.A., Matzner-Löber, E. Nonparametric estimation of the conditional mode with errors-in-variables: strong consistency for mixing processes. *J. Nonparametr. Stat.*, **14** (3), 341-352 (2002).

[JO97]   Jacob, P., Oliveira, P. Kernel estimators of general Radon-Nikodym derivatives. *Statistics*, **30**, 25-46 (1997).

[JH01]   James, G., Hastie, T. Functional linear discriminant analysis for irregularly sampled curves. *J. R. Stat. Soc. Ser. B* **63**, 533-550 (2001).

[JHS00] James, G., Hastie, T., Sugar, C. Principal component models for sparse functional data. *Biometrika*, **87**, 587-602 (2000).

[JS05]   James, G., Silverman, B. Functional adaptive model estimation. *J. Amer. Statist. Assoc.* , **100**, 565-576 (2005).

[JS03]   James, G., Sugar, C. Clustering for sparsely sampled functional data. *J. Amer. Statist. Assoc.* **98**, 397-408 (2003).

[JMVS95] Janssen, P., Marron, J.S., Veraverbeke, N. Sarle, W. Scale measures for bandwidth selection. *J. Nonparametr. Statist.*, **5**, 359-380 (1995).

[J84]   Jirina, M. On regular conditional probabilities. *Czechoslovak Math. J.*, **9**, 445-451 (1984).

[K87]   Kemperman, J.H.B. The median of a finite measure on a Banach space. In: *Statistical data analysis based on the L1-norm and related methods*, North-Holland, Amsterdam, 217-230, (1987).

[K94]   Kim, T.Y. Moment bounds for non-stationnary dependent sequences. *J. Appl. Probab.*, **31** (3), 731-742 (1994).

[KD02]   Kim, T.Y., Kim, D. Kernel regression estimation under dependence. *J. Korean Statist. Soc.*, **31** (3), 359-368 (2002).

[KSK04] Kim, T.Y., Song, G.M., Kim, J.H. Robust regression smoothing for dependent observations. *Commun. Korean Math. Soc.*, **19** (2), 345-354 (2004).

[KG92]   Kneip, A., Gasser, T. Statistical tools to analyze data representing a sample of curves. *Ann. Statist.* **20** 1266-1305 (1992).

[KLMR00] Kneip, A., Li, X., MacGibbon, K. B., Ramsay, J. O. Curve registration by local regression. *Canad. J. Statist.*, **28**, 19-29 (2000).

[KB78]   Koenker, R., Bassett, G. Regression quantiles. *Econometrica*, **46**, 33-50 (1978).

[KO05]   Krause, A., Olson, M. *The Basics of S-PLUS*. Springer-Verlag, New York, 4th Edition (2005)

[K86]   Krzyzak, A. The rates of convergence of kernel regression estimates and classification rules. *IEEE Trans. Inform. Theory*, **32**, 668-679 (1986).

[K91]   Krzyzak, A. On exponential bounds on the Bayes risk of the kernel classification rule. *IEEE Trans. Inform. Theory*, **37**, 490-499 (1991).

[L56]   Lanczos, C. *Applied analysis*. Prentice Hall, Inc., Englewood Cliffs, N. J. (1956).

[LFR04] Leao, D., Fragoso, M., Ruffino, P. Regular conditional probability, integration of probability and Radon spaces. *Proyecciones*, **23**, 15-29 (2004).

[L89]   Lehtinen, M., Paivarinta, L., Somersalo, E. Linear inverse problems for generalised random variables. *Inverse problems*, **5**, 599-612 (1989).

[L99]   Liebscher, E. Asymptotic normality of nonparametric estimators under α-mixing condition. *Statistics and Probability Letters* **43** (3), 243-250 (1999).

[LS72]   Lipster, R., Shiryayev, A. On the absolute continuity of measures corresponding to processes of diffusion type relative to a Wiener measure. (in Russian). *IZv Akad. Nauk. SSSR Ser. Math*, **36** (1972).

[LS01]   Li, W.V., Shao, Q.M. Gaussian processes: inequalities, small ball probabilities and applications. In : C.R. Rao and D. Shanbhag (eds.) *Stochastic processes, Theory and Methods*. Handbook of Statistics, **19**, North-Holland, Amsterdam (2001).

[LMS93] Leurgans, S.E., Moyeed, R.A., Silverman, B.W. Canonical correlation analysis when the data are curves. *J. R. Statist. Soc. B*, **55**, 725-740 (1993).

[LMSTZC99] Locantore, N., Marron, J., Simpson, D., Tripoli, N., Zhang, J., Cohen, K. Robust principal component analysis for functional data (with discussion). *Test*, **8**, 1-74 (1999).

[LO99]   Louani, D., Ould-Sad, E. Asymptotic normality of kernel estimators of the conditional mode under strong mixing hypothesis. *J. Nonparametr. Statist.*, **11** (4), 413-442.

[L96]   Lu, Z.D. Consistency of nonparametric kernel regression under α-mixing dependence. (in Chinese) *Kexue Tongbao*, **41** (24), 2219-2221 (1996).

[LP94]   Lugosi, G., Pawlak, M. On the posterior-probability estimate of the error rate of nonparametric classification rule. *IEEE Trans. Inform. Theory*, **40**, 475-481 (1994).

[M85]   Ma, Z.M. Some results on regular conditional probabilities. *Acta Math. Sinica (N.S.)*, **1**, 302-307 (1985).

[MBHG96] Macintosh, A., Bookstein, F., Haxby, J., Grady, C. Spatial pattern analysis of functional brain images using partial least squares, *Neuroimage*, **3**, 143-157 (1996).

[MR03]   Malfait, N., Ramsay, J. The historical functional linear model. *Canad. J. Statist.*, **31**, 115-128 (2003).

[M93]   Mammen, E. Bootstrap and wild bootstrap for high dimensional linear models. *Ann. of Statist.*, **21**, 255-285 (1993).

[M95]   Mammen, E. On qualitative smoothness of kernel density estimates. *Statistics*, **26**, 253-267 (1995).

[M00]   Mammen, E. Resampling methods for nonparametric regression. In: M. Schimek (Ed) *Smoothing and regression; Approaches, Computation, and Application. Wiley Series in Probability and Statistics*, Wiley, New York, 425-451 (2000).

[MN89]   Marron, J.S., Nolan, D. Canonical kernels for density estimation. *Statistics and Probability Letters* **7**, 195-199 (1989).

[MNa89]   Martens, H., Naes, T. *Multivariate Calibration*. Wiley, New York (1989).

[ME99]   Marx, B., Eilers, P. Generalized linear regression on sampled signals and curves: a P-spline approach. *Technometrics*, 41, 1-13 (1999).

[M89]   Masry, E. Nonparametric estimation of conditional probability densities and expectations of stationary processes: strong consistency and rates. *Stochastic Process. Appl.*, **32** (1), 109-127 (1989).

[M05]   Masry, E. Nonparametric regression estimation for dependent functional data: asymptotic normality. *Stochastic Process. Appl.*, **115** (1), 155-177 (2005).

[M65]   Massy, W.F. Principal Components Regression in Exploratory Statistical Research. *J. Amer. Statist. Assoc.* **60**, 234-246 (1965).

[M00]   Mint El Mouvid M. Sur l'estimateur linéaire local de la fonction de répartition conditionnelle (in french). Phd, Université Montpellier 2, (2000).

[MM03]   Mugdadi, A., Munthali, E. Relative efficiency in kernel estimation of the distribution function. *J. Statist. Res.* **37**, 203-218 (2003).

[N64]   Nadaraya, E. On estimating regression. *Theory Proba. Appl.* **10**, 186-196 (1964).

[N76]   Nadaraya, E. Nonparametric estimation of the Bayesian risk in a classification problem. *Sakharth. SSR Mecn. Akad. Moambe* (in Russian), **82**, 277-280 (1976).

[N97]   Nagaev, S. Some refinements of probabilistic and moment inequalities. *Teor: Veroyatnost. i Primenen* (in russian), **42** (4), 832-838 (1997).

[N98]   Nagaev, S. Some refinements of probabilistic and moment inequalities (English translation). *Theory Probab. Appl.*, **42** (4), 707-713 (1998).

[NCA04]   Naya, S., Cao, R., Artiaga, R. Nonparametric regression with functional data for polymer classification. In *Proceedings in Computational Statistics*, ed. J. Antoch, Physica-Verlag, Heidelberg New York, (2004).

[NN04]   Nazarov, A., Nikitin, Y. Exact L_2 small ball behavior of integrated Gaussian processes and spectral asymptotics of boundary value problems. *Probab. Theory Related Fields*, **129** (4), 469-494 (2004).

[O60]   Obhukov, A. The statistically orthogonal expansion of empirical functions. *American Geophysical Union*, 288-291 (1960).

[O97]   Ould-Saïd, E. A note on ergodic processes prediction via estimation of the conditional mode function. *Scand. J Statist.*, **24**, 231-239 (1997).

[PT01]   Pagès, G., Tenenhaus, M. Multiple factor analysis combined with PLS path modelling. Application to the analysis of relationships between physico-chemical variables, sensory profiles and hedonic judgements, *Chemometrics and Intelligent Laboratory Systems*, **58**, 261-273 (2001).

[P62]   Parzen, E. On estimation of a probability density function and mode. *Ann. Math. Statist.*, **33**, 1065-1076 (1962).

[P88]   Pawlak, M. On the asymptotic properties of the smoothed estimators of the classification error rate. *Pattern Recognition*, **21**, 515-524 (1988).

[PSV96] Pélégrina, L., Sarda, P., Vieu, P. : On multidimensional nonparametric regression estimation. In: A. Prat (ed.) *Compstat 1996*, Physica-Verlag, 149-162 (1996).

[P93]   Pesin, Y.B. On rigourous mathematical definitions of correlation dimension and generalized spectrum for dimensions. *J. Statist. Phys.*, **71**, 529-573 (1993).

[PT98]   Poiraud, S., Thomas, C. Quantiles conditionnels (in french). *J. Soc. Franc. de Statist.*, **139**, 31-44 (1998).

[P06]   Preda., C. Regression models for functional data by reproducing kernel Hilbert space methods. *J. Statist. Plann. Inference*, to appear (2006).

[PS05]   Preda, C., Saporta, G. PLS regression on a stochastic process. *Comput. Statist. Data Anal.*, **48**, 149-158 (2005).

[PS05b] Preda, C., Saporta, G. Clusterwise PLS regression on a stochastic process. *Comput. Statist. Data Anal.*, **49**, 99-108 (2005).

[QV97]   Quintela-Del-Ro, A., Vieu, P. A nonparametric conditional mode estimate. *J. Nonparametr. Statist.*, **8** (3), 253-266 (1997).

[RDCT]   R Development Core Team. A language and environment for statistical computing. R Foundation for Statistical Computing, Vienna, Austria (2004).

[RV05]   Rachdi, M., Vieu, P. Sélection automatique du paramètre de lissage pour l'estimation nonparamétrique de la régression pour des données fonctionnelles (in french). *Compte Rendus Acad. Sci. Paris*, Ser. I, **341**, 365-368 (2005).

[RV05b] Rachdi, M., Vieu, P. Nonparametric regression for functional data: automatic smoothing parameter selection. Preprint (2005).

[RT97]   Ray, B., Tsay, R. Bandwidth selection for kernel regression with long-range dependent errors. *Biometrika*, **84** (4), 791-802 (1997).

[RS97]   Ramsay, J., Silverman, B.W. *Functional Data Analysis*. Springer-Verlag, New York (1997).

[RS02]   Ramsay, J., Silverman, B.W. *Applied functional data analysis; Methods and case studies*. Springer-Verlag, New York (2002).

[RS05]   Ramsay, J., Silverman, B.W. *Functional Data Analysis. 2nd Edition*. Springer-Verlag, New York (2005).

[R81]   Reiss, R.-D. Nonparametric estimation of smooth distribution functions. *Scand. J. Statist.*, **8**, 116-119 (1981).

[R00]   Rio, E. *Théorie asymptotique des processus aléatoires faiblement dépendants* (in french). Springer, Mathématiques & Applications, **31**, (2000).

[R83]   Robinson, P. Nonparametric estimators for time series. *J. Time Ser. Anal.* **3**, 185-207 (1983).

[R97]   Robinson, P. Large-sample inference for nonparametric regression with dependent errors. *Ann. Statist.*, **25** (5), 2054-2083 (1997).

[R56]   Rosenblatt, M. A central limit theorem and a strong mixing condition. *Proc. Nat. Ac. Sc. U.S.A.* **42**, 43-47 (1956).

[R69]   Rosenblatt, M. Conditional probability density and regression estimators. In: P.R. Krishnaiah (ed.) *Multivariate Analysis II*, Academic Press, New York (1969).

[Ro69]   Roussas, G. Nonparametric estimation of the transition distribution function of a Markov process. *Annals of Mathematical Statistics* , **40** 1386-1400 (1969).

[Ro90]   Roussas, G. Nonparametric regression estimation under mixing conditions. *Stochastic Process. Appl.*, **36** (1), 107-116 (1990).

[Ro96]   Roussas, G. Exponential probability inequalities with some applications. In: *Statistics, probability and game theory.* IMS Lecture Notes Monogr. Ser., Inst. Math. Statist., Hayward, CA. **30**, 303-319 (1996).

[RI87]   Roussas, G., Ioannidès, D. Moment inequalities for mixing sequences of random variables. *Stoch. proc. and Appl.*, **5** (1), 61-120 (1987).

[S89]   Samanta, M. Non-parametric estimation of conditional quantiles. *Statist. Proba. Letters*, **7**, 407-412 (1989).

[SM90]   Samanta, M., Thavaneswaran, A. Nonparametric estimation of the conditional mode. *Comm. Statist. Theory Methods*, **19**, 4515-4524 (1990).

[SV00]   Sarda, P., Vieu, P. Kernel regression. In: M. Schimek (ed.) *Smoothing and regression; Approaches, Computation, and Application.* Wiley Series in Probability and Statistics, Wiley, New York, 43-70 (2000).

[S00]   Schimek, M. (ed.): *Smoothing and regression; Approaches, Computation, and Application.* Wiley Series in Probability and Statistics, Wiley, New York (2000).

[S81]   Schumaker, M. *Spline functions: basic theory.* Wiley (1981).

[S92]   Scott, D. *Multivariate density estimation. Theory, practice, and visualization.* Wiley Series in Probability and Mathematical Statistics. John Wiley & Sons, New York (1992).

[S96]   Shi, Z. Small ball probabilities for a Wiener process under weighted supnorms, with an application to the supremum of Bessel local times. *J. Theoret. Probab.*, **9**, 915-929 (1996).

[S03]   Shmileva, E. Small ball probabilities for a centered Poisson process of high intensity. (in russian) *Zap. Nauchn. Sem. S.-Petersburg. Otdel. Mat. Inst. Steklov.*, **298** (6) , 280-303 (2003.

[S95]   Silverman, B. W. Incorporating parametric effects into functional principal components analysis. *J. R. Stat. Soc. Ser. B* **57** 673-689 (1995).

[S96]   Silverman, B. W. Smoothed functional principal components analysis by choice of norm. *Ann. Statist.*, **24**, 1-24 (1996).

[S01]   Skld, M. The asymptotic variance of the continuous-time kernel estimator with applications to bandwidth selection. *Stat. Inference Stoch. Process*, **4** (1), 99-117 (2001).

[S82]   Stone, C. Optimal global rates of convergences for nonparametric estimators. *Ann. Statist.* **10**, 1040-1053 (1982).

[S85]   Stone, C. Additive regression and other nonparametric models. *Ann. Statist.* **13**, 689-705 (1985).

[TK03]   Tarpey, T., Kinateder, K. Clustering functional data. *J. Classification*, **20**, 93-114 (2003).

252   References

[T94]   Tjöstheim, D. Non-linear time series: a selective review. *Scand. J. Statist.*, **21** (2), 97-130 (1994).

[TA94]   Tjöstheim, D., Auestad, B. Nonparametric identification of nonlinear time series: selecting significant lags. *J. Amer. Statist. Assoc.*, **89** (428), 1410-1419 (1994).

[T63]   Tortrat, A. *Calcul des Probabilités* (in french). Masson, Paris (1963).

[U37]   Uspensky, J. *Introduction to mathematical probability.* McGraw-Hill, New York (1937).

[VR00]   Venables, W., Ripley, B. *S Programming.* Springer, New York (2000).

[V91]   Vieu, P Quadratic errors for nonparametric estimates under dependence. *J. Multivariate Anal.*, **39** (2), 324-347 (1991).

[V93]   Vieu, P. Bandwidth selection for kernel regression: a survey. In: W. Härdle and L. Simar (eds.) *Computer Intensive methods in Statistics.* Statistics and Computing, Physica Verlag, Berlin, 134-149 (1993).

[V95]   Vieu, P. Order choice in nonlinear autoregressive models. *Statistics*, **26** (4), 307-328 (1995).

[V96]   Vieu, P. A note on density mode estimation. *Statist. Probab. Lett.*, **26**, 297-307 (1996).

[V02]   Vieu, P. Data-driven model choice in multivariate nonparametric regression. *Statistics*, **36** (3), 231-246 (2002).

[W90]   Wahba, G. *Spline models for observational data.* SIAM, Philadelphia (1990).

[WZ99]   Wang, H.J., Zhao, Y.S. A kernel estimator for conditional t-quantiles for mixing samples and its strong uniform convergence. (in chinese). *Math. Appl. (Wuhan)*, **12** (3), 123-127 (1999).

[W64]   Watson, G. Smooth regression analysis. *Sankhya Ser. A*, **26**, 359-372 (1964).

[W66]   Wold, H. Estimation of principal components and related models by iterative least squares. In: P. Krishnaiaah (ed.) *Multivariate Analysis.* Academic Press, New York, 391-420 (1966).

[W75]   Wold, H. Soft modelling by latent variables; the non linear iterative partial least squares approach. In: J. Gani (ed.) *Perspectives in Probability and Statistics, Papers in Honour of M.S. Bartlett.* Academic Press, London (1975).

[X94]   Xue, L.G. Strong consistency of double kernel estimates for a conditional density under dependent sampling. (in chinese) *J. Math. (Wuhan)*, **14** (4), 503-513 (1994).

[YMW05] Yao, F., Müller, H.-G., Wang, J.-L. Functional data analysis for sparse longitudinal data. *J. Amer. Statist. Assoc.* , **100**, 577-590 (2005).

[Y92]   Yoshihara, K. *Weakly dependent stochastic sequences and their applications. I: Summation theory for weakly dependent sequences.* Sanseido (1992).

[Y93a]   Yoshihara, K. *Weakly dependent stochastic sequences and their applications. II: Asymptotic statistics based on weakly dependent data.* Sanseido (1993).

[Y93b]   Yoshihara, K. *Weakly dependent stochastic sequences and their applications. III: Order statistics based on weakly dependent data.* Sanseido (1993).

[Y93c]   Yoshihara, K. Bahadur-type representation of sample conditional quantiles based on weakly dependent data. *Yokohama Math. J.*, **41** (1), 51-66 (1993).

[Y94]   Yoshihara, K. *Weakly dependent stochastic sequences and their applications. IV: Curve estimation based on weakly dependent data.* Sanseido (1994).

[Y94b]   Yoshihara, K. *Weakly dependent stochastic sequences and their applications. V: Estimators based on time series.* Sanseido (1994).

[Y96]   Youndjé, E. Propriétés de convergence de l'estimateur à noyau de la densité conditionnelle (in french). *Rev. Roumaine Math. Pures Appl.*, **41**, 535-566 (1996).

[YSV94] Youndjé, E., Sarda, P., Vieu, P. Validation croisée pour l'estimation non-paramétrique de la densité conditionnelle. (in french). *Publ. Inst. Statist. Univ. Paris*, **38**, 57-80 (1994).

[YJ98]   Yu, K., Jones, M.C. Local linear quantile regression. *J. Amer. Statist. Assoc.*, **93**, 228-237 (1998).

[Y76]   Yurinskii, V. Exponential inequalities for sums of random vectors. *J. of Multiv. Analysis*, **6**, 475-499 (1976).

[ZL85]   Zhao, L., Liu, Z. Strong consistency of the kernel estimators of conditional density function. *Acta Math. Sinica (N.S.)*, **1**, 314-331 (1985).

[ZL03]   Zhou, Y., Liang, H. Asymptotic properties for L_1 norm kernel estimator of conditional median under dependence. *J. Nonparametr. Stat.*, **15** (2), 205-219 (2003).

Index

Springer Series in Statistics *(continued from p. ii)*

Huet/Bouvier/Poursat/Jolivet: Statistical Tools for Nonlinear Regression: A Practical
Guide with S-PLUS and R Examples, 2nd edition.
Ibrahim/Chen/Sinha: Bayesian Survival Analysis.
Jolliffe: Principal Component Analysis, 2nd edition.
Knottnerus: Sample Survey Theory: Some Pythagorean Perspectives.
Kolen/Brennan: Test Equating: Methods and Practices.
Kotz/Johnson (Eds.): Breakthroughs in Statistics Volume I.
Kotz/Johnson (Eds.): Breakthroughs in Statistics Volume II.
Kotz/Johnson (Eds.): Breakthroughs in Statistics Volume III.
Küchler/Sørensen: Exponential Families of Stochastic Processes.
Kutoyants: Statistical Influence for Ergodic Diffusion Processes.
Lahiri: Resampling Methods for Dependent Data.
Le Cam: Asymptotic Methods in Statistical Decision Theory.
Le Cam/Yang: Asymptotics in Statistics: Some Basic Concepts, 2nd edition.
Liu: Monte Carlo Strategies in Scientific Computing.
Longford: Models for Uncertainty in Educational Testing.
Manski: Partial Identification of Probability Distributions.
Mielke/Berry: Permutation Methods: A Distance Function Approach.
Molenberghs/Verbeke: Models for Discrete Longitudinal Data.
Nelsen: An Introduction to Copulas, 2nd Edition.
Pan/Fang: Growth Cure Models and Statistical Diagnostics.
Parzen/Tanabe/Kitagawa: Selected Papers of Hirotugu Akaike.
Politis/Romano/Wolf: Subsampling.
Ramsay/Silverman: Applied Functional Data Analysis: Methods and Case Studies.
Ramsay/Silverman: Functional Data Analysis, 2nd edition.
Rao/Toutenburg: Linear Models: Least Squares and Alternatives.
Reinsel: Elements of Multivariate Time Series Analysis, 2nd edition.
Rosenbaum: Observational Studies, 2nd edition.
Rosenblatt: Gaussian and Non-Gaussian Linear Time Series and Random Fields.
Särndal/Swensson/Wretman: Model Assisted Survey Sampling.
Santner/Williams/Notz: The Design and Analysis of Computer Experiments.
Schervish: Theory of Statistics.
Seneta: Non-negative Matrices and Markov Chains, Revised Printing.
Shao/Tu: The Jackknife and Bootstrap.
Simonoff: Smoothing Methods in Statistics.
Singpurwalla and Wilson: Statistical Methods in Software Engineering: Reliability and
Risk.
Small: The Statistical Theory of Shape.
Sprott: Statistical Inference in Science.
Stein: Interpolation of Spatial Data: Some Theory for Kriging.
Taniguchi/Kakizawa: Asymptotic Theory of Statistical Inference for Time Series.
Tanner: Tools for Statistical Inference: Methods for the Exploration of Posterior
Distributions and Likelihood Functions, 3rd edition.
van der Laan: Unified Methods for Censored Longitudinal Data and Causality.
van der Vaart/Wellner: Weak Convergence and Empirical Processes: With Applications
to Statistics.
Verbeke/Molenberghs: Linear Mixed Models for Longitudinal Data.
Weerahandi: Exact Statistical Methods for Data Analysis.
West/Harrison: Bayesian Forecasting and Dynamic Models, 2nd edition.

 **Springer** **springeronline.com**
the language of science

Applied Functional Data Analysis
Methods and Case Studies
J.O. Ramsay and B.W. Silverman

The authors' highly acclaimed book *Functional Data Analysis* (1997) presented a thematic approach to the statistical analysis of such data. The present book introduces and explores the ideas of functional data analysis by the consideration of a number of case studies, many of them presented for the first time. The two books are complementary but neither is a prerequisite for the other. The case studies are accessible to research workers in a wide range of disciplines.

2002. 190 p. (Springer Series in Statistics) Softcover ISBN 0-387-95414-7

Functional Data Analysis
Second Edition
J.O. Ramsay and B.W. Silverman

Scientists and others today often collect samples of curves and other functional observations. This monograph presents many ideas and techniques for such data. This second edition is aimed at a wider range of readers, and especially those who would like to apply these techniques to their research problems. There is an extended coverage of data smoothing and other matters arising in the preliminaries to a functional data analysis. The chapters on the functional linear model and modeling of the dynamics of systems through the use of differential equations and principal differential analysis have been completely rewritten and extended to include new developments. Other chapters have been revised substantially, often to give more weight to examples and practical considerations.

2005. 436 p. (Springer Series in Statistics) Hardcover ISBN 0-387-40080-X

All of Nonparametric Statistics
Larry Wasserman

The goal of this text is to provide the reader with a single book where they can find a brief account of many, modern topics in nonparametric inference. This text covers a wide range of topics including: the bootstrap, the nonparametric delta method, nonparametric regression, density estimation, orthogonal function methods, minimax estimation, nonparametric confidence sets, and wavelets. The book has a mixture of methods and theory.

2005. 276 p. (Springer Texts in Statistics) Hardcover ISBN 0-387-25145-6

Easy Ways to Order▶ Call: Toll-Free 1-800-SPRINGER • E-mail: orders-ny@springer.sbm.com • Write: Springer, Dept. S8113, PO Box 2485, Secaucus, NJ 07096-2485 • Visit: Your local scientific bookstore or urge your librarian to order.